The Essential Book of Fermentation

The Essential Book of Fermentation

Great Taste and Good Health
with Probiotic Foods

JEFF COX

With Illustrations by the Author

AVERY

a member of Penguin Group (USA) Inc.

New York

Published by the Penguin Group
Penguin Group (USA) Inc., 375 Hudson Street,
New York, New York 10014, USA

USA · Canada · UK · Ireland · Australia
New Zealand · India · South Africa · China

Penguin Books Ltd, Registered Offices: 80 Strand, London WC2R 0RL, England
For more information about the Penguin Group visit penguin.com

Most Avery books are available at special quantity discounts for bulk purchase for sales promotions, premiums,
fund-raising, and educational needs. Special books or book excerpts also can be created to fit specific needs.
For details, write Penguin Group (USA) Inc. Special Markets, 375 Hudson Street, New York, NY 10014.

Library of Congress Cataloging-in-Publication Data

Cox, Jeff, date.
The essential book of fermentation : great taste and good health with probiotic foods / Jeff Cox.
pages   cm
Includes bibliographical references and index.
ISBN 978-1-58333-503-1
1. Fermentation.   2. Fermented foods.   3. Probiotics.   I. Title.
QR151.C65   2013   2013009623
572'.49—dc23

Printed in the United States of America
3 5 7 9 10 8 6 4 2

BOOK DESIGN BY TANYA MAIBORODA

ALWAYS LEARNING   PEARSON

For Elizabeth, Allison, Sean, Isis, and Aria

Plant root hairs (left) and intestinal villi (right) have a large surface area that greatly increases nutrient absorption capacity from the soil and intestinal contents, respectively.

Contents

Preface:
An Epiphany About Health

One day, a scientific paper crossed my desk at *Organic Gardening* magazine describing how certain plots of land produced scabby potatoes—caused by a soil-borne fungus—while other fields were scab-free. In an experiment, soil from the healthy field was mixed into the scab-producing soil, and that immediately cured the problem. The reason, the scientists reported, was that a strong mix of microorganisms in the healthy soil literally ate up the fungus that produced scab. In the soil, it turns out, possession is nine-tenths of the law, and soils colonized by overwhelmingly large numbers of beneficial bacteria and other good microbes just don't leave much room for disease-causing organisms to get a toehold.

Not only that, but I discovered that a wide range of soil microorganisms inhibit or kill other microbes—especially pathogenic ones that cause rot and disease in plants. This piqued my interest because it was fungus control naturally, without chemical fungicides, and fit right in with the organic method. At the same time, I was teaching myself soil science by reading Nyle Brady's standard textbook on the subject, *The Nature and Properties of Soils*. Brady was writing about how soil microbes are intimately involved in the decomposition of organic matter. "In addition to their direct attack on plant tissue, they are active within the digestive tracts of some animals," he wrote.

Well, sure, I thought. Fruits and vegetables, green leaves, nuts, and other organic matter are not only digested by the microbes in the compost pile and in the soil, but also by the microorganisms that inhabit every healthy human intestine and actually are ubiquitous all over the planet. In fact, all the plants and animals in the world weigh about the same as the world's total microbial biomass. I did some further reading on the subject and was astonished to find that nine out of every ten cells in our bodies are intestinal microorganisms, and that those cells contain 99 percent of the DNA in our bodies. They live in our gut and decompose organic matter exactly the same way they do in the compost pile. When scientists looked at a healthy human intestinal flora, they found more than four hundred species of bacteria, thirty to forty of which compose about 99 percent of the flora. (Now we know that every gram of our intestinal contents contains somewhere in the neighborhood of a trillion bacteria, and that the gastrointestinal tract is the most densely colonized region of the human body.)

Longtime organic gardeners know that good, healthy compost contains similar numbers of microorganisms in each spoonful, and that rich organic soil is even more diverse than our gut bacteria, with somewhere in the neighborhood of ten thousand species of microbes. By one estimate, there may be 150 million species of microbes in the world, almost all as yet unclassified. These myriad bacteria and fungi are at the very root of health—of the soil, of plants, and of animal and human health. Maybe, I thought, a healthy mix of the proper intestinal microorganisms builds our natural good health the way it does in the soil.

This was an exciting idea. I found out several things that supported it. Our intestinal flora is indeed a source of good health. Intestinal bacteria such as *Lactobacillus acidophilus* not only decompose our food, rendering it into forms our intestines can absorb to feed our bodies, but in doing so they themselves actually manufacture vitamins such as vitamin K and many of the B vitamins—including B_{12}. Further, they thrive in the acid conditions of our digestive tract, hence their species name (*acidophilus* means "acid-loving"). Disease-causing organisms prefer pH neutral conditions and have a hard time getting established in acid conditions. I asked a nutritional researcher which foods most strongly promote the establishment of a healthy intestinal flora, and he said, "If half your diet is fresh, raw fruits and vegetables, the other half can be Twinkies and you'll be healthy." In other words, the same thing that supports the proper nutrition of plants—actively decaying organic matter—is the same thing that supports the proper nutrition of human beings.

Then I had an epiphany—one of those visions that throws such a strong light on a subject that it changes the way you look at the world forever. I thought of a plant root and how at the small tips of the root, microscopic root hairs extend out into the soil to absorb the nutrients that microorganisms produce there. And then I thought of a human intestine and how it's lined with microscopic villi—tiny projections much like root hairs—that point inward into the intestine and absorb nutrients from the decomposing elements of last night's dinner.

I got it! An intestine is a root turned inside out! It carries its soil within it. This is the adaptation that eons ago separated animals from plants. It's as if animals learned how to pull their roots out of the soil and turn them inside out so that they could walk around instead of being anchored to one spot.

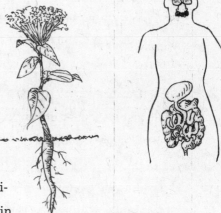

This was an important insight for me, both in my job communicating the value of organic culture to a million readers and to me personally, as it showed me the way forward to a healthy diet. Practicing organic agriculture and eating fermented foods both result in health—organic agriculture yields a healthy farm ecosystem, and fermented food yields health in the gut ecosystem. The same processes are at work, with the same and similar microorganisms in many cases giving the same result: healthy food and healthy people.

The key here is the concept of an ecosystem: a network or web of interconnected and interdependent life forms that support one another. The study of ecology is a relatively new science, with many fathers and mothers, perhaps none as important as Howard Thomas Odum and his brother Eugene, who first synthesized many loosely affiliated ideas into the study of ecosystem ecology, which then developed into the science of ecology. Of course, pieces of the puzzle had been formulated as long ago as ancient Greece, where Aristotle and Theophrastus studied animals and their relationships. Darwin's ideas were an important advance, and many others developed insights that finally clicked into place in the 1960s when the culture at large became concerned with the environment, and the science of ecology was born in its modern form.

The key concept is that everything is in some way connected to everything else, or as Francis Thompson, the British poet, put it:

All things by immortal power,
Near and Far
Hiddenly
To each other linked are,
That thou canst not stir a flower
Without troubling of a star.

This transcendental thought is not only undergirded by the study of ecology, but also by quantum mechanics, where two discrete atomic particles can affect each other simultaneously at a distance, with no apparent connection between them.

A second key concept is that when it comes to the health of an ecosystem, the more varied the life forms in it, the healthier it is, and that health is defined by the system's stability; that is, its resistance to abrupt dislocation and change, as by a disease sweeping through, or by the sudden population explosion of a single species that swamps other members of the ecosystem. Rabbits overran Australia because there were no foxes or other predators to keep them in check. The phylloxera louse that attacks the roots of European varieties of grapevines nearly wiped out the French vineyards in the late nineteenth century, which were saved only when phylloxera-resistant native American grape rootstock was rushed to France and the European vines were grafted on it. Deadly lionfish escaped their aquariums in Florida not long ago, and now are found in the Atlantic all along the East Coast of the United States, again because their natural predators did not colonize the Atlantic with them. In a healthy ecosystem, there are adequate numbers of predators and prey, making the system stable. In any indigenous ecosystem, predators and prey have coevolved. The prey are a trophic niche that calls forth a predator to fill it. The ultimate prey animal is the plant-eating rodent—mouse, shrew, vole, rabbit, gopher, and many others—and predators from raptors to foxes to cats use rodents for food. This great diversity helps the stability of the ecosystem in which these animals are found. If an epidemic of disease knocks out an area's foxes, raptors and cats will take up the slack. Diversity means that natural backup systems are in place, if and when needed.

A third concept is that ecosystem diversity is enhanced by recycling of the system's nutrients by the fermentative process. In the wild, this happens naturally as each year's tree leaves and herbaceous plants decay into the soil. In a forest managed for timber or firewood, cutting down the trees and carting them away without returning nutrients to the soil will eventually produce depleted soil that's prone to erosion and wind damage and becomes increasingly devoid of life. The need for recycling is even more urgent on a farm. One of the big problems with conventional farms is that little waste is recycled. Soil becomes simply a means of propping up crops. Fertilizers are applied as soluble nitrogen, phosphorus, and potassium compounds made in factories.

On the organic farm, however, farm wastes and manures are composted. What nutrients and elements are taken off the land and sold as farm products are more than replaced by the sheer proliferating biomass of microbes produced in the composting process. Finished compost has much more fertilizing power than the raw ingredients that went into the compost pile.

And so, in the human gut, the nutrients that are eliminated from your body in bowel movements are more than compensated for by the healthy ecosystem that digests the raw materials (your food). The gut is an internalized composting facility, feeding you exactly what nature intends for your health, when you need it, and in the forms that will do you the most good. Like any healthy ecosystem, the presence of a diverse mix of bacteria and yeasts is stable and disease-resistant, and in fact goes further in its protective function by bolstering your immune system.

Look to the organic garden for a clear outward depiction of what goes on inside you. There are no "bad" insects and "good" insects—only a healthy mix of plant-eating insects and insect-eating insects. Garden plants actually grow better when nibbled by insects because the nibbling stimulates the production of plant hormones, which support better growth and greater production of fruits and seeds. And the nibblers provide food for the insect eaters, and so the system is balanced and healthy. Now if you poison the insects, it's the insect-eating bugs that are most susceptible to the pesticides. These chemicals reduce the system's diversity in a selective way, killing off the most susceptible insects first. This makes it much more likely for plant eaters to proliferate and cause problems. And so it is with the use of antibiotics. If you are really being overwhelmed by an infection, antibiotics can be a lifesaver. But the routine use of antibiotics or their presence in your food will damage your intestinal ecosystem, making it much more likely for one or another pathogen to proliferate in your gut and cause further illness. The best way to prevent that is to maintain healthy gut ecology to begin with, or, if you must take an antibiotic, drink kefir and kombucha, eat fermented vegetables, and replenish the gut dwellers as fast as the antibiotics can kill them off.

These correspondences are manifold. The web of life is complicated in that picking that flower does indeed trouble that star. Consequences are hard to predict because the system is so complex. But that doesn't mean it's fruitless to try.

Let's take a look at a rigorous and peer-reviewed scientific study to see if organic agriculture really works, and if it's validated, what that might mean for our ingestion of probiotic foods. One such study was done in 2010 on a multi-university level that included private and public teams of scientists.[*] They found that the organic farms had strawberries with longer shelf life, greater dry matter (meaning that when the water was removed, there was more substance to the berries), and higher antioxidant activity and concentrations of ascorbic acid (vitamin C) and phenolic compounds, but lower concentrations of phosphorus and potassium. "In one variety," the scientists reported, "sensory panels judged organic strawberries to be sweeter and have better flavor, overall acceptance, and appearance than their conventional counterparts." The organic farm soils had more biodiversity, took more nitrogen from the air and fed it to the plants, and degraded residual chemical pesticides; that is, the soils were healthier.

Here are the paper's conclusions in its own words:

"Our findings show that the organic strawberry farms produced higher quality fruit and that their higher quality soils may have greater microbial functional capability and resilience to stress."

You'll have to forgive me a personal note here. After listening to apologists for the chemical agriculture industry deride organics for the last forty years, after hearing that half the population would have to starve if we went organic, and that there are no studies that show organic crops are in any way superior to those grown conventionally, I must say that this one study—joined by many, many others—should silence the critics. But of course it won't. Press releases from agribusiness flacks and hacks continue.

The "strawberry fields" study implies something else: that just as a diversity of microbes is important in promoting positive health, so is diversity in foodstuffs. Our government suggests we eat at least three to five different kinds of fruits and vegetables a day. Many European countries suggest from seven to nine of these foods, and in Japan, citizens are advised to eat eleven or more different foods each day. It's not the quantities that count, it's the diversity of foods that produce a well-

---

[*] J. P. Reganold, P. K. Andrews, J. R. Reeve, L. Carpenter-Boggs, C. W. Schadt, et al., "Fruit and Soil Quality of Organic and Conventional Strawberry Agroecosystems," *PLOS ONE* 5, no. 2 (2010): e12346. doi:10.1371/journal.pone.0012346.

rounded diet. The food diversity supplies many different nutrients and substances that not only feed us directly, but that also stimulate a diversity of microbes in the gut, with each tasked to disassemble a different food.

Just as compost—a veritable seething pile of microscopic life—is the beating heart that feeds nutrients to the healthy organic garden, so is our intestine an internalized compost pile that feeds nutrients to our bodies.

The point of this book is that the same natural energies, processes, laws, tendencies, and holistic approaches to food production that reveal the superiority of organic agriculture are applicable to nutrition as augmented by fermented foods. In both cases, it's working with nature's ecosystems of microbes that produces the amazing results.

Our body is like a garden that needs organic care to thrive. That means avoiding toxic chemicals in our food, but it also means avoiding antibacterial soap when we wash. Scientists today are finding that the steep rise in autoimmune diseases among children may be due in large part to the fact that kids today spend a lot more time indoors with their computers and video games, are washed with antibacterial soaps, have their hands wiped frequently with hand sanitizer, and are watched over by parents who are concerned about ill effects from germs.

But this approach is backward. The scientists now say that it's a lack of exposure to a wide variety of microbes that leads to compromised and underdeveloped immune systems; that the immune system needs lots of contact with germs to develop a diverse immune response. There was a time, not long ago, when kids got plenty dirty playing outside or working and helping around the family farm. The world outside is absolutely alive with microbes everywhere. In the organic garden, microbes are cherished, put to work recycling nutrients; they are like a flame in the soil, with the gardener heaping more fuel on this living fire with every shovelful of compost. The same approach to what we eat translates into a world of good health for us, and the way to that result is through the ingestion of fermented and fermenting foods.

And so our health is in some measure predicated on the health of our gut ecology, just as plant health is predicated on the health of soil ecology. It seemed right then and still seems right that nature builds the health of her advanced children (large plants and animals) on the strong ecological health of her smallest creatures: the fungi, yeasts, bacteria, actinomycetes, and other microorganisms that decompose organic matter. They are, after all, the destiny and source of all life.

Shakespeare, of course, said it all centuries ago when the friar in *Romeo and Juliet* remarks:

> What's Nature's mother is her tomb.
> What is her burying grave, that is her womb.

The Essential Book
of Fermentation

Introduction

Within these regions, battles rage; populations rise and fall; affected just as we are by local environmental conditions. Industry thrives and constant defense is exercised against interlopers and dangerous aliens who may enter unannounced. Colonists roam and settle—some permanently, some only briefly. In general we have in miniature many of terrestrial life's vicissitudes, problems, and solutions.

—LEON CHAITOW AND NATASHA TRENEV, *Probiotics*

What is fermentation? It's the process whereby microbes turn the stuff of one generation of plants and animals into food for the next generation of plants and animals. They do this by decomposing once living matter into its constituent nutrients to be taken up by living creatures, but also by enriching the plant and animal matter with their own bodies and metabolites.

Maybe a more appropriate question for us as humans is, Why not fermentation? As a way to preserve food and improve its nutritional qualities, the technique has been around for thousands of years. Actually, it's been around for billions of years in wild nature. You can't stop it. Milk curdles into cheese, grape juice turns into wine, and a mixture of flour and water rises—all on their own. Or, rather, by the action of unseen microbes.

When it comes to microbes, we've generally thrown the baby out with the bathwater. We pasteurize, sterilize, and can food by killing all the microbes that could spoil the food. These processes eliminate the microbes that spoil food, but also kill the microbes that improve the taste and nutrition of that food. My decades of experience learning about, writing about, and practicing organic gardening have convinced me that our lives are inextricably bound up with the lives of microbes, and that when their health is factored into our decisions, we all thrive.

You will find, as you read this book, that I often make reference to organic agriculture and gardening as a way to understand fermentation. Through my long association with organics, I've come to see profound similarities between what happens on the organic farm and in the organic garden and what happens in the fermenting vessel.

Canning food to preserve it is all about killing microbes and disease-causing organisms. Fermentation may involve some processes that resemble canning techniques, but it is a fundamentally different approach to food that encourages microbial growth rather than trying to squelch it. Conventional farming and gardening is all about killing, too, using pesticides, herbicides, fungicides, and antibiotics. Organic culture, on the other hand, is all about stimulating diversity—the realization that good health in the garden or on the farm is a result of a strong and diverse web of life in the soil. The very word "health" comes from an Anglo-Saxon root meaning "whole," and the whole system most definitely includes the microbes. Fermenting food for its health benefits is all about promoting a diverse ecosystem, too, only one that exists within us. Organic gardeners and farmers know that health is built from the ground up. They understand that if you feed the soil, the soil will feed the plants. This book extends that insight to say: Feed your intestinal flora and your intestinal flora will feed you.

The trick is to know how to produce fermented foods that are good for you, while preventing the growth of organisms that are bad for you. That's relatively simple, as you'll find out in this book.

First of all, knowing how to safely ferment foods is fun. It's easy to find others who have this knowledge and share this passion (start online at www.fermenters club.com). It's geeky and it's cool. But above all, it's tasty and it's good for you.

How good? Very, very good, in some surprising ways. The microbes that ferment foods not only make raw ingredients taste better and become more nutritious, they participate in the healthy web of life that lives within us and on us and coats every surface we touch. They float in the air and enter us with each breath we take, every swallow we make, and each bite of food we eat. And when they are happy, so are we. Many of the most delicious foods are fermented. Think of bread, cheese, and wine as examples. All are fermented and together they make a fine meal indeed.

More than that, fermentation involves fundamental processes that are at the heart of life itself in all its diverse functions. When leaves fall in the forest, they ferment and mold into soil-building humus. Yesterday's leaves feed today's trees, which produce tomorrow's leaves, and so life begets life.

In nature, the whole is contained in every part, like fractals, which are shapes that can be repeatedly divided into parts that are smaller copies of the original shape. Think of the way water swirls down a drain and the arms of a spiral galaxy. A natural law or tendency is repeated through all creation, from the microscopic to the macroscopic. The laws and tendencies that create health in the soil of an organic garden or farm apply equally to human health, and to the health of the wild places, and even the health of the planet as a whole, and it has to do with the life that's invisible to us because it's too small—the world of one-celled microbes. We can see their effects in the bubbling of a fermenting tank of wine or in the blackening and steaming ingredients of a well-made compost pile, but they are out of our range of sight.

These laws begin with the simple recognition that all life must nourish itself. There are a few forms of life that find nourishment in inorganic chemicals—one thinks of the tube worms feeding on the ammonia and methane that erupt from volcanic vents in the deep sea floor—but almost all life feeds on the animal or vegetable flesh of other life forms that have decomposed through the fermenting action of microbes.

The greater the number and kinds of microbes in a soil, the healthier that soil is and the more able it is to promote the health of the plants and animals that live from that soil. That's because in a natural ecosystem, all the available environmental and food niches tend to be filled. The organisms that fill those niches coevolved over millions of years, and their lives are tied together like parts in a complicated machine. When a natural ecosystem reaches its climax state, it is sustainable in perpetuity as long as the environment doesn't change. It functions to its maximum. In other words, the more diverse the ecosystem, the more closely it approaches its climax state, giving the soil more fermentation power than a state where the soil is disturbed and damaged by the application of toxic chemicals. The more fermentation power in the soil, the more dead plant and animal matter will be thoroughly decomposed into its nutritive elements, and the plants that grow there will consequently be better fed and healthier. The more fermentation power in the guts of animals, the healthier those animals will be. And that includes us. And as we saw in the preface to this book, the more biodiversity in the soil, the harder it will be for disease organisms to break out and do damage.

That insight spawned the modern organic movement when Sir Albert Howard published *An Agricultural Testament* in 1940. Howard made the discovery that cattle grazing on compost-fertilized pasture avoided diseases that afflicted cattle grazing on land fertilized with factory-made chemical compounds—fertilizers that did nothing to add actively decaying organic matter to the soil. The key to soil health and the health of the life that derives from it, he said, was this actively decaying organic matter, or, as we identify the essential process in this book, fermentation.

In 1960, most fermented foods found in the United States were ethnic specialties, such as the sauerkraut found among the German enclaves of the Pennsylvania Dutch farmers in Pennsylvania. If you wanted to eat organic food, you pretty much had to grow it yourself, and if you wanted to eat fermented organic food, you pretty much had to ferment it yourself. But by 1970, a huge shift in public awareness had occurred. The National Environmental Policy Act had recently passed. The first Earth Day was held on April 22 of that year. *The Greening of America,* by Charles Reich, would soon top the charts of best-sellers. The back-to-the-land movement took off, aided by the digest-sized magazine *Organic Gardening & Farming.* From 225,000 subscribers in 1970, the magazine reached 1.2 million subscribers in 1980, and the folks who put out the magazine threw themselves a big party when it finally, after forty years, turned a profit. Fortunately for me, I was in the middle of all this as the magazine's managing editor.

If I was going to write intelligently about organic gardening techniques and back-to-the-land skills, I realized that I needed to learn them. We back-to-the-landers also did some fermenting, learning to make beer and wine, cheese, tofu, and sauerkraut, and learning that farming was hard work and serious business, requiring knowledge of a wide range of disciplines such as botany, soil science, entomology, ecology, plant pathology, meteorology, and more, all overseen by a knowledge of how to turn a profit. Most of us eventually scurried back to the city, where we bought business attire and engaged for the next twenty-five to thirty years in making money and raising families. I was doubly fortunate in that my paying job was to preach what I practiced at home on my two acres in the deep country.

And now the great wheel of time has turned some more, and the helix of life is welcoming a new generation of people back to nature. This time the entry point is less pie in the sky and more actual pie on the plate. Here is nature's grand scheme writ small and doable. An ever-increasing number of people are finding that nature works her wonders in the kitchen and in the basement, in jars and tubs and crocks, and that the very air is charged with sparkling bits of life looking for raw materials to turn into edible gold.

You can build your own health one delicious swallow at a time, strengthening your immune system, working with nature to enhance the quality of your food. There's little you have to buy to achieve this, but quite a lot to learn. Hence this book.

PART 1

The What and Why of Fermentation

In this section, we discover the world of microbes and how fermentation is one of their most central functions for the support and enhancement of life. We see how we are connected to all of nature through the operation of microbes and cells. We gain a new perspective on the health of nature and the nature of our own health. And we discover the amazing roles played by microbes within us and on us.

CHAPTER 1

Fermentation— The Engine of Life

Nature uses the nutrients from past life to create new life for the future using the technique of decomposition.

With the fermented foods we humans enjoy so much, we catch the process of decomposition while it is still new and fresh and wholesome, at its most delicious, nutritious, and healthful for us. Not only are the ingredients in our fermented foods being opened up so their nutrients are more easily available to us, but the bloom of microbes in the fermented foods provide added nutrients as their bodies finish their work and spill their nutritional cell contents into our intestines. They do this at a furious rate. One bacterium can split into two bacteria and the two can split into four, and so on, so that within twenty-four hours, there can be millions and billions of bacteria from just that one cell alone. Many of these bacterial cells take a beneficial place within our digestive tracts for more or less time, depending on the type of microorganism supporting our digestive function where the process of decomposition into usable nutrients is completed.

As we make our fermented foods, let's realize that the process we are working with is the very engine of life. Pay attention to its transformative power and imbibe its goodness with thanks. An example of this transformative power is that fermented foods improve digestion. That's because fermentation is, in one sense, the predigestion of food, arrested as the food becomes more nutritious and better tasting, but before the digestion proceeds to decomposition. People who are lactose intolerant, for instance, can sometimes eat yogurt. That's because lactose, the compound that people can't tolerate, is predigested by enzymes produced by microbes in the yogurt. The action of the microbes that cause predigestion is arrested by the refrigerator. The cold conditions in a fridge don't stop the microbes completely—freezing does that—but it slows them enough for our food to stay safe and tasty much longer than if it were sitting out on the counter. The microbes that make kefir, for example, work twice as fast at 85°F as they do at 70°F.

As fermentation progresses, it becomes central to the decomposition of organic matter. Decomposition is nature's janitorial function, ridding the world of old organic matter through the action of microorganisms. Without this process, we'd be slogging through millions of years of dead plants and animals piled up way over our heads. Fermentation is one of the factors in decomposition, but it's also something central to how life works. It is nature's all-important method of recycling nutrients through plants and animals. However, it works in tiny ways, so small that the fermentation actors are out of our sight. But given the impact of this infinitesimal world on our health and the health of the planet, these actors are worth a closer look.

If we could reduce ourselves in size by a factor of a thousand, we could see microbes and discover that they coat everything, inside and out. They are everywhere. For-

Without decomposition, we'd be slogging through millions of years of dead plants and animals piled way up over our heads.

est soil rich in organic matter is fairly made from them. They cover our bodies and the bodies of every other creature on earth, plant and animal. They are a part of us. They make up 90 percent of the cells and more than 99 percent of the genes and DNA in our bodies. "Don't assume microbes are simple," says Dimitar Sasselov, an astrobiologist at the Harvard-Smithsonian Center for Astrophysics. Paul Falkowski, a biophysicist and biologist from Rutgers, adds, "Animals are overgrown microbes. We are here to ferry microbes around the planet. Plants and animals are an afterthought of microbes."

"We are here to ferry microbes around the planet."

The invisible inhabitants of the microbial world rule this planet. They have been here for billions of years and are exquisitely and successfully adapted to life on earth. Try as we might to wipe them out with antibiotics, antimicrobial gels, and bactericidal soaps, all we succeed in doing is selecting for resistant strains that quickly reproduce, presenting us with super-strong versions of themselves. It's the same reason that pesticides don't work. They may curtail various insects' activities in the short run, but soon new and improved versions of those insects will come storming back, able to ignore those pesticides. It's nature's way. She has strength in numbers, and evolves resilient populations of microbes that overcome whatever we humans, in our shortsighted attempts at control, throw at them.

As on the organic farm, the best strategy is to work with nature, not against her. If nature wins, we all win. If nature loses to our biocidal and aggressive tactics, we all lose. And so we welcome the bacteria and yeasts of nature's invisible world into our lives. We understand what her little minions need and what they do that benefits us.

While we can't see this minuscule world directly with the naked eye, we can see the results they produce as easily as we see the invisible wind fluttering through a field of wheat or flapping laundry on a clothesline. A vat of wine grape juice seems to boil with bubbling billows of carbon dioxide as yeast consumes sugar. A dish of flour and water set on a kitchen counter soon becomes alive with microorganisms that will happily leaven our bread. The carcass of a dead animal slowly disappears into the environment—eaten by vultures, decomposed by microbes, its bones scattered by hungry dogs or covered with a blanket of forest leaves that eventually turn to soil that buries the evidence.

If we could see the microbial world, we'd also notice that these specks of life form colonies and that the colonies form mutually beneficial or antagonistic ecologies. And these ecosystems follow the same laws that govern the ecosystems here on our macro level. As ecosystems mature, they become more diverse. As they become more diverse, they build a healthy, interconnected web of life among the participants. Eventually they reach what's called a climax ecosystem, which changes slowly as the participants evolve. A climax ecosystem has reached its destiny and, if not disturbed, is the definition of sustainable. This is as true of a microbial ecosystem as of an old-growth forest ecosystem.

Without microbial life, the plants and animals we know in our macro world would not—could not—exist, for there would be no way for them to get nourishment. With no microorganisms to turn sunlight into sugar, to turn the nitrogen gas in the atmosphere into nitrate fertilizers, to dismantle dead organic matter and turn it into life-giving nutrients, life as we know it would be impossible.

When microbial ecosystems are diverse, strong, and healthy, the plants and animals in our macro world that depend on them partake of that health, too. Our human-level ecosystems flourish when we stop the destruction of life caused by industrial and agricultural chemicals. Health is built from the ground up—literally—and fermentation has a huge role to play in a more sensible future.

Imagine what would happen if we as a society mimicked nature's role in turning dead organic matter into plant food. Every year in the United States, many billions of tons of clean organic waste are mixed with toxic wastes and indigestible plastics, glass, and metals and dumped into landfills. We call it waste, and its potential is certainly wasted because there is a world of nutrition in that organic waste that could be recovered. Imagine if municipalities had a collection day for clean, organic waste—no toxics and no indigestibles allowed. If this waste was

ON THE HEAD OF A PIN

In the Middle Ages, theologians argued over how many angels could dance on the head of a pin. Today, we fermenters might consider how many bacteria and yeast cells could fit on that pinhead.

Figure a pinhead is about two millimeters in diameter. A micrometer is one thousandth of a millimeter, so creatures a few micrometers in size are too small to be seen with the naked eye. An average bacterium is about two micrometers long and half a micrometer wide, while yeast cells are circular and about six micrometers in diameter. So the answer to our question is that about 10,000 bacteria or 3,000 to 4,000 yeast cells can fit on the head of a pin.

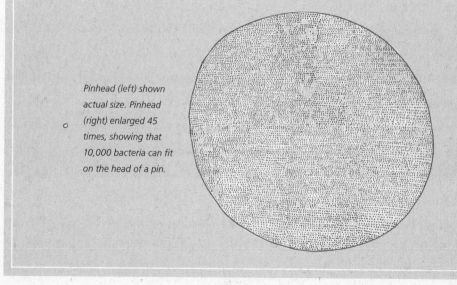

Pinhead (left) shown actual size. Pinhead (right) enlarged 45 times, showing that 10,000 bacteria can fit on the head of a pin.

fermented by microorganisms, it would enormously enrich the waste. Enzymes change starches like paper, wood, and garbage into sugars that produce alcohol when fermented. In the process, the waste becomes enriched with the bodies of the cells doing the fermenting—it's more nutritious coming out of the fermentation than going in. Sound familiar? The resulting mass of digested waste would be high-class fertilizer. Instead of being buried in landfills where it putrefies and contaminates groundwater, the waste could enrich the soil on the farms in the food sheds surrounding our cities and the alcohol could help power our vehicles or be burned to power generators that make electricity. Even now, fermenting garbage and organic waste in the Puente Hills landfill near Los Angeles produces landfill gas, mostly methane, that's being captured and burned to boil water to make steam to drive turbines that produce 46 megawatts of electricity that's sold to Southern California Edison, enough to meet the power needs of nearly 70,000 homes. While this is commendable, controlling the fermentation in large vessels could yield a wide range of useful products along with electricity.

We are in the vanguard of those who know the potential of fermenting, not only in our personal lives through the probiotics we ingest, but in the potential enhancement of the health of the farmland that grows our food and supports our own health even further. This is good, environmentally sound, sustainable, organic, and positive thinking that needs translation into action. We are the people who can push to bring it about.

Let's take a quick look at the overall process of decomposition of dead plant and animal tissue. The tiniest single-celled microbes, like any living thing, need to eat. And what they like to eat is organic matter; that is, anything that was once alive or, in the case of some microbes that cause infections, that is still alive. As microbes decompose organic matter, they tear apart old cells and turn rigid cell walls and cell contents that are often in unusable form into liquid nutrients that seep into the soil, where living plant roots absorb them and build new tissue. Where plants grow and die, season after season, the soil is enriched by this recycling method. Where plants are eaten by animals, the same process of decomposition happens, only it happens internally in the gut and the nutrients are recycled into bone and muscle and blood and all the rest of the animal's organs. If a carnivore eats an herbivore, the nutrients are recycled once again into the tissues of the carnivore's body. And when the carnivore dies, its body is once again recycled back into its constituent nutrients by microbes, and the wheel of life continues to turn.

That's the big picture. Nature, in her intricate way, has assigned types of microbes different roles in this process of decomposition and recycling. Plant nutrients are classified as macronutrients—elements that plants need in quantity to grow—and micronutrients, no less important for healthy growth but needed in far fewer quantities. The macronutrients are nitrogen, phosphorus, and potassium. The following are some of the chief types of microbes assigned to these nutrients and their duties.

The Nitrogen Recyclers

Some microbes are the recyclers of nitrogen in what soil scientists call the nitrogen cycle, turning yesterday's plants and animals into food for tomorrow's plants and animals. All life depends on their work dismantling old organic matter. Microbiologists know about tens of thousands of these bacteria and other microbes, but there may be many thousands and maybe millions as yet undiscovered and named. Most of them fall into five overall functional classifications:

1. Living tissue contains proteins that are composed of nitrogen-containing molecules. When tissue dies, certain microbes use the carbon and water in it as their food, while its nitrogen is a waste product of their metabolism that is converted to ammonium salts and then into ammonia. **Nitrifying bacteria** grab the nitrogen from the ammonia and link it with oxygen atoms to form soluble nitrates that plants absorb through their roots to make new living tissue.

2. **Nitrogen-fixing bacteria** colonize certain plants' roots. They take molecules of nitrogen from the air (N_2) and split them apart by rearranging their electrons, like taking apart a Chinese wire puzzle. These freestanding atoms of nitrogen are then reassembled by the bacteria with oxygen atoms to form soluble nitrates, which plants absorb to make new living tissue. The end result is the same as with nitrifying bacteria but by a different process.

3. **Denitrifying bacteria** are usually anaerobic, meaning they operate without the presence of air. They convert the nitrogen in dead plant tissue to nitrogen gas molecules (N_2) that are given off into the air. Remember that the air we breathe is four-fifths nitrogen, some of which is recycled back into plant nutrients by the nitrogen-fixing bacteria. Instead of aerobic decomposition, anaerobes cause putrefaction—the reason putrid piles of wet plants smell.

4. **Actinobacteria** are aerobic microbes that are central to the process of decomposing dead plants' nitrogen-rich amino acids into ammonium salts, which are further converted to ammonia. Their activity turns dead plant tissue into sweet-smelling humus, which is a critical component of healthy soil.

5. **Fungi** are also important in decomposing organic matter into nitrates that living plants can use as food. A succession of fungal species colonizes organic matter during its decomposition, beginning with those that decompose sugars and starches, succeeded by those that are able to break down the tough materials of cellulose and lignin, releasing the nitrogen from their proteins.

The Phosphorus Scavengers

A special type of fungus called mycorrhizal fungi has a symbiotic relationship with its host plants. It lives in the soil and colonizes the roots of plants. Their plant hosts exude sweet syrup through their roots that the fungus uses for food. In return, the fungus sends long, threadlike strings called hyphae far into the surrounding soil to gather phosphorus and transport it back to the plant roots, where it's absorbed. Phosphorus, a macronutrient essential for plant health, is often in short supply in many soils, so you can see the valuable service mycorrhizal fungi play in the natural ecosystem. If the plants are harvested and removed, then the phosphorus will eventually be depleted in that soil and must be added back as fertilizer. If the plants are growing wild, the phosphorus will be recycled back into the soil through decomposition.

A bacillus is a rod-shaped bacterium (left).
A coccus is round (right).

The Carbon Managers

Then there is the genus *Lactobacillus*, which includes the microbes at the heart of the fermentations that do so much for our health. They turn sugars into lactic acid, an acid that links carbon with hydrogen and oxygen to form the molecules of adenosine triphosphate (ATP) in our bodies that provide us with energy as it is broken down into its constituent pieces. We have only about eighty-five grams (about three ounces) of ATP in our bodies at any one time, and when exercising we'd soon use it all up, but the body is a marvelous piece of work and has several systems for resynthesizing ATP from its breakdown products. One of those systems depends on lactic acid, and lactobacilli move the carbon from sugars to lactic acid in our intestines, so you can see their importance. This genus is composed of at least 120 species, with possibly more as yet undiscovered.

The lactobacilli operate under a variety of conditions. When working under the brine in our fermentations, they are anaerobes that don't need air to function. Yet they also work in the presence of air, where they play a major role in the decomposition of plant organic matter. They are rod-shaped eukaryotes; that is, they have nuclei inside their cell walls that contain their genetic material. On the human body, they are found primarily in the digestive tract and in the vagina. Lucky that they live there, too, for the vagina is the birth canal, and baby's trip out into the world seeds its little gut with these benign, essential bacteria. But lactobacilli presence in the vagina is good for women, too. They maintain balance in the vaginal ecosystem, and they protect the lining of the vagina by producing a thick layer of cells that are a barrier to pathogens—including *Candida albicans*, the pesky yeast that causes yeast infections. They maintain the vagina's pH at an acidic 4.5. And they generate H_2O_2—hydrogen peroxide—which carries that extra oxygen atom to a pathogen or parasite, where the oxygen combines with and kills the offender. They protect against various kinds of pathogens. They are symbiotic with their human hosts in both these places, living off the human body and in turn giving the human beneficial substances like lactic acid and performing a wide range of health-promoting activities.

Of course, nobody's perfect, and that includes the lactobacilli. A high count of lactobacilli in the mouth has been a standard test for dental caries for many years. It seems that they promote the progression of tooth decay by turning sugar into lactic acid—normally a good thing, but when it happens in the plaque clinging to your teeth, the lactic acid tends to dissolve the minerals in the teeth, leading to cavities. So brush your teeth, and don't forget to floss. But forget the antibacterial mouthwashes, which have the same effect on the healthy bacteria in the mouth as fungicides do on soil microorganisms. A potent strain of *Streptococcus salivarius*, found in the mouths of healthy humans, is called BLIS K12, and it fights pathogens that cause bad breath, among other maladies. Antibacterial mouthwashes wreak wholesale havoc on bacterial ecologies in the mouth and kill off S. *salivarius* and beneficial lactobacilli.

Besides lowering pH and inhibiting pathogens, the lactobacilli can reduce inflammation and fight cancer. A 2009 study by researchers at Beth Israel Deaconess Medical Center and UCLA showed that some strains of these bacteria help the body fight off cancer and prevent tumor formation. Animal studies have shown inhibition of liver, colon, bladder, and mammary tumors when lactobacilli were added to the animals' diet.

For the chemists among us (and you know who you are), here's the reaction where glucose is reduced to lactic acid in a process called homofermentation (because only one compound is produced):

$$C_6H_{12}O_6 \rightarrow 2\ CH_3CHOHCOOH$$

Some strains of lactobacilli produce what's called a heterofermentation, because more than one compound is produced. Here's a formula for the reduction of glucose into one molecule of lactic acid, one molecule of ethanol, and one molecule of carbon dioxide:

$$C_6H_{12}O_6 \rightarrow CH_3CHOHCOOH + C_2H_5OH + CO_2$$

In both cases, the lactobacilli turn the sugar molecule into either one or two molecules of lactic acid. And as we just found out, lactic acid is part of one of the ways the body resynthesizes ATP to power our systems.

The primary mechanism of ATP production is a reduction of glucose in the blood into pyruvate, a process called glycolysis that is facilitated by lactic acid, which acts like an enzyme. One of the results of glycolysis is the release of hydrogen ions (protons). It's not the lactic acid that causes the pain in muscles when they are worked really hard, but rather the buildup of protons in the cells, causing the pH of the cells to drop and turn acidic.

ATP production in the cells happens by an electron transport chain. Just picture a molecule of glucose coming apart into pyruvate molecules, and energy jumping from electron to electron, producing ATP's precursors, and finally, with the help of lactic acid, producing ATP itself. To power the body's cells in their many functions—working muscles, active digestive system, firing of nerve cells—the cells disassemble ATP back into its precursors, and the lactic acid is there to reassemble it for further energy. In effect, in the anaerobic conditions inside the cell, glucose is effectively undergoing lactic acid fermentation.

If we turn now to the organic garden or farm, we find that plants, too, use ATP to energize their systems. During the production of ATP in plant cells, hydrogen atoms (each is a proton and an electron) are stripped of their electrons, leaving positively charged protons—and in effect, the synthesis of ATP in root hairs creates proton pumps that send protons into the soil solution, as groundwater is called. Proton pumps are simply membranes that enclose the cell's nucleus and mitochondrion and form the whole cell's walls. They collect protons and store them to be used as an energy source in a wide range of cell functions in both plants and animals. Plants seeking nutrients release stored protons through the cell walls of root hairs into the groundwater. These protons find their way to humus particles in the soil. Humus is what's left after the decomposition of once living tissue is completed. Humus is marvelous stuff, and soils handled organically are full of it. Its surfaces are negatively charged and hold positively charged ions of plant nutrients to themselves: ammonium, calcium, magnesium, and potassium, among others, which prevents these plant foods from being washed away by rain or irrigation.

As the protons from the plant root hairs reach the humus particles, they kick the plant nutrients loose from the humus and replace them at the humus surface attachment sites. The plant nutrients then float back to the plant roots and feed the plants. This function is called the Cation Exchange Capacity of a soil, and the more humus, the more energetic the soil's Cation Exchange Capacity. This is one reason organic soils have it all over soils fertilized with chemicals. The latter are soluble and wash away, where they can cause environmental damage due to flooding groundwater with overloads of soluble nutrients.

Why do I bring this up in a book on fermentation and human nutrition? Because we also have proton pumps in our gastrointestinal system, especially in our stomachs. In our stomachs, these positively charged protons find negatively charged ions of chlorine (Cl-), derived mostly from NaCl (salt), and link up to form HCl—hydrochloric acid, which is our gastric juice and a strong acid that is one of the first stages of digesting the food we swallow. So proton pumps have similar functions in plants and in us—they aid in our ability to feed ourselves.

Fermented foods can help balance the production of hydrochloric acid. Too little—sometimes a consequence of advancing age—reduces the stomach's ability to give food a bath in strong acid that begins to reduce the food to its nutritional constituents. Fermented foods can increase the acidity of the stomach's gastric juices because of the lactic acid they contain. But they act as a buffer if the stomach overproduces this very strong acid (pH 2.0). Kefir has a pH of about 4.0, and sauerkraut about 3.0. As the fermented foods pass through the stomach, they can help protect the intestinal lining from the hydrochloric acid in the gastric juice.

Your digestive system requires enzymes to properly break down your food and make full use of it. Enzymes are catalysts and allow for the full utilization of the nutrients in food. They are destroyed by heat hotter than 120 to 130°F—especially the heat of pasteurization. If you eat a cooked potato that has had its enzymes destroyed in the cooking, your body rushes white blood cells to your intestines. These cells contain digestive enzymes that set to work deconstructing the potato. However, if you eat just a small amount of raw potato, that you might grate into a salad, for instance, the enzymes will be intact and your immune system won't have to rush white blood cells to your intestine. But who wants to eat raw potato? The answer is to eat fermented foods, because most are not cooked and are full of digestive enzymes. So instead of the body's having to rush white blood cells to your intestines to digest the cooked potato, your intestinal flora chips right in and helps with the deconstruction. Researchers have found that as we age, the supply of digestive enzymes produced by the body diminishes. Fermented foods keep us fully supplied.

We've seen that vitamin C in fermenting cabbage is more available to the body than the vitamin in raw cabbage. But this good result doesn't stop with vitamin C. Fermented dairy products increase the level of folic acid, so critical in the healthy development of the fetus. Kefir in particular increases the level of riboflavin and biotin, two B vitamins, depending on the strains of bacteria present.

Finally, fermenting our foods makes them taste better, or at least more complex and intense. Many people feel that wine tastes better than grape juice, bread tastes better than flour, cheese tastes better than plain milk, and sauerkraut tastes better than raw cabbage, to name just a few examples. Our taste buds have developed over the millennia to guide us toward foods that are good for us. Newborns like a sweet taste because the mother's milk that is their lifeline tastes sweet. When we were hunter-gatherers—through 99 percent of our species' time on earth—salt, sugar, and fat were in short supply, and so we crave those tastes. That gets us into trouble today, when foods rich in salt, sugar, and fat—doughnuts, pizza, ice cream— are readily available. Now we need to exhibit restraint when it comes to these tastes. But you know what? Fermenting vegetables, milk, and other foods twists their flavor profile to satisfy these primordial cravings, even if the foods are not rich in salt, sugar, and fat. For example, grape juice tastes sweet because when grapes are ripe, about 25 percent of their weight is grape sugar. Yeast ferments that sugar into about half that weight in alcohol, which doesn't taste sweet. And yet a well-made, rich red wine, like a Napa Valley Cabernet Sauvignon, strikes many people as sweet, even though there's very little sugar in it. What they are tasting is the intense fruitiness of the wine, which only makes it seem sweet.

A Personal Health Story

Over a hundred years ago, Elie Metchnikoff, a Russian biologist, zoologist, and protozoologist who's best remembered today for his pioneering research into the immune system, received the Nobel Prize in Medicine. Among his many lines of fertile inquiry, he found that kefir activates the flow of saliva, most likely due to its lactic acid content and its slight amount of carbonation. He also found that kefir stimulates peristalsis and digestive juices in the intestinal tract.

His work was strictly scientific, but I would like to add a personal anecdotal note about the effect of drinking kefir daily on the intestinal tract and the eliminative function. I would, if I could, recount my experience delicately, but perhaps it's best told plainly, since we all share the same bodily functions—and, sometimes, dysfunctions.

I noticed, as I aged, that the strength of peristalsis—the squeezing function of the colon in moving feces toward elimination—began to taper off. More and more, I had to bear down to achieve a positive bowel movement. This bearing down by contracting the stomach muscles is dangerous. One can pop a blood vessel doing that. And it often resulted in bothersome hemorrhoids that at least itched when they weren't positively painful. Not only that, but the more I used my stomach muscles to help peristalsis, the weaker peristalsis seemed to get and the harder I had to push, making the problem worse.

I had no idea that drinking kefir would help with this problem. I was surprised by the results of drinking my daily morning glass of kefir, for within a week, my bowel movements were larger and the new bulk stimulated my colon to regain its peristaltic oomph. Within a few weeks I could relax and let nature take its course; that is, let the body pursue elimination naturally and easily, without pushing. I would say the stool bulk increased from 30 to 50 percent after I started drinking kefir. The bacteria and yeast were obviously proliferating like crazy in my intestines, and it was all to the good.

Often there would be a second productive bowel movement later in the morning, with the same positive results. Hemorrhoids were gone. I felt clean and healthy inside and out. Finding kefir has made a significant difference in the quality of my personal hygiene. Sorry if it's an indelicate subject, but I thought you'd like to know, because you might find the same benefits from a daily glass of delicious kefir.

Some mornings I drink the kefir plain—it's that good. But I also went to the market and bought a bag of frozen wild Maine blueberries, a bag of frozen dark sweet organic cherries, and a bag of frozen organic black raspberries. At home, I dumped them into a pot, set the pot on low heat with the lid on, and when the fruit was thawed and hot, mashed it with a potato masher. Then I lined a colander with three layers of cheesecloth, sat it on a bowl, and poured the contents of the pot into the cheesecloth. I worked the solids back and forth with a tablespoon until all the juice had collected in the bowl. Then I stirred in two tablespoons of Sucanat and the juice of a lemon, poured this mixture into an empty quart jar, and put it in the fridge. On some mornings when I want a treat, I pour a few tablespoons of this syrupy, fruity mix into my glass of kefir and stir it up. Delicious!

This last delightful information is to make up for all the bathroom talk that preceded it.

CHAPTER 2

Fermented Foods as Probiotics

The long lives of some Bulgarian and Russian citizens have been credited to their consumption of large amounts of yogurt. The key organism that turned milk into yogurt was later identified as *Lactobacillus acidophilus*. This by-now-familiar organism (see the milk department of any supermarket), at work in yogurt, cheese, and other dairy products, enzymatically changes lactose—milk sugar—into lactic acid. It may do this in a dish of ripening yogurt or in the human intestine. There are some major benefits of this change. The lactobacilli attach themselves temporarily to sites all over the digestive tract, reducing the places where disease-causing germs can gain a foothold. They help the existing intestinal microbes absorb and metabolize the lion's share of nutrients, making them unavailable to pathogens. They produce lactic and acetic acids, creating an acid environment unsuitable for harmful organisms, which prefer conditions near pH neutral. They turn lactose into lactic acid, reducing the discomfort of those who are lactose intolerant. Research has documented the ability of lactic acid–producing bacteria to inhibit *Escherichia coli*, *Salmonella typhimurium*, *Staphylococcus aureus*, and *Clostridium perfringens*—all disease-causing pathogens. Studies have shown that these activities also stimulate the immune system.

In fact, recent scientific work in Brazil shows that our diverse intestinal micro-biome is responsible for developing immune system responses that lead to inflammation and pain; that is, they help our immune systems acquire the ability to cause swelling and pain. This ability protects us, claims Arthur Ouwehand, a microbiologist at the University of Turku in Finland. He points out that pain and swelling are the body's correct responses to a wound or infection. They signal the brain that something is wrong and that the immune system is working on the problem. The pain makes the person treat the injured area more carefully. The Brazilian studies were led by Mauro Teixeira, an immunologist who found that the presence of a healthy gut biome stimulated the immune systems of mice, producing larger numbers and more active white blood cells, among other benefits. The results imply that the right mix of diverse microorganisms in the gut could relieve pain in people with inflammatory diseases such as Crohn's disease and eczema by healing the damaged gut, but such therapies are far from certain at this stage, Teixeira says. Still, his research "shows how profound an effect microbes have on your immune system and your entire health," Ouwehand says.

That's not all. Lactic acid bacteria produce bacteriocins. Bacteriocins can be likened to natural antibiotics that clear an area of competition, allowing the lactobacilli, in this case, to have the territory to themselves. They are attack proteins produced by many kinds of bacteria, not only lactobacilli. Most have a very broad operating range and remain active even after being frozen or boiled. They work by dissolving a hole in the cell wall of the competing bacteria, causing the cell contents to leak out and the cell to die. Because bacteriocins are a naturally occurring protein, they are assimilated back into the host after they finish their task.

If you think of yourself as a bacterium, and know that everywhere you want to get a toehold is already crammed with your competition, you can see why nature has given you the ability to spread your surroundings with a bacteriocide. It gives you a chance. The most thoroughly studied bacteriocin is nisin, produced by our friends so intimately involved in fermentation, the lactobacilli. It's an approved food additive in the European Union and is used as a preservative in many kinds of cheese. Nisin is produced from the two main lactic acid–producing bacteria and protects cheddar, Colby, Monterey Jack, cottage cheese, brick, Limburger, Muenster, Swiss, Emmental, Gouda, Edam, Brie, Camembert, blue, Gorgonzola, Stilton, feta, Manchego, chèvre, and many others. It works like any other allelopathic bacteriocin: preventing other microbes from getting established.

German researchers are now investigating the possibility of using bacteriocins as a substitute for antibiotics to eliminate resistance caused by the overuse of antibiotics in humans and animals. Research has revealed that antibiotics actually *increase* the colonization of an intestinal tract by disease-causing salmonella bacteria, whereas the bacteriocins produced by lactic acid bacteria inhibit or eliminate it. Well, how can that be? Why would antibiotics increase the colonization of the intestinal tract by salmonella?

Organic gardeners and farmers know the answer. After a field is sterilized with chemicals that destroy most of its life forms, the first organisms to recolonize the field will be opportunistic weeds and plant-eating pestiferous insects that eat those weeds. The first plants back will certainly not be climax ecosystem plants that take many generations to appear as the ecosystem becomes diverse and stable. And the only insects that can thrive in a disturbed field recolonized by opportunistic weeds are insects that can eat those weeds. Only after the plant-eating insects become dominant will the good guys—the insect-eating insects—return. Evidently, the same idea holds true for an intestine cleansed of bacteria by an antibiotic. The first organisms back in may not be the good guys, unless steps are taken to make sure the first organisms back in are members of a healthy intestinal floral ecosystem. And that means seeding your sterilized intestines with microbe-rich fermented foods.

Fermented foods stimulate the body's production of acetylcholine, a neurotransmitter that helps nerve impulses fire through their neurons. What this means is that functions like bowel peristalsis—the movement of feces through the bowel—are aided by acetylcholine, and this can improve regularity and curb constipation. It also helps the stomach, pancreas, and gallbladder produce their digestive juices and enzymes, which can greatly aid digestion.

The fact that fermented foods are predigested by bacteria lightens the load on the pancreas as it produces insulin that metabolizes carbohydrates. Carbohydrates in fermented foods are already broken down to an extent by the action of lactobacilli. People with diabetes, whose pancreases aren't producing enough insulin, may benefit from consumption of ferments.

Fermented foods destroy and inhibit pathogenic bacteria. These foods protect us from harmful bacteria in many ways, by herding pathogens into colonies that are then isolated, by flooding the intestines with beneficial bacteria so there's little room for pathogens to break out, by creating an acid environment that pathogens don't like, by producing bacteriocins that control and inhibit disease-causing germs, and by other ways still not understood.

Dr. David Williams, a chiropractor and authority on natural healing techniques, says that the idea of using beneficial microbes to control pathogenic microbes is just catching on, "although it's still baffling to me why conventional medicine hasn't been able to grasp or accept this concept. Obviously, pharmaceutical companies stand to make a lot more money selling antibiotics and other medications than by recommending a daily dose of fermented cabbage." Here's another example of the correspondence between conventional medicine and conventional farming. Let's repeat the last statement, substituting a few words: The idea of using beneficial insects to control plant-eating pests is just catching on, although it's still baffling to me why conventional agriculture hasn't been able to grasp or accept this concept. Obviously, agricultural chemical companies stand to make a lot more money selling pesticides and other agrichemicals than by recommending the annual release of ladybugs.

Conventional medicine has too long been in thrall to the big pharmaceutical manufacturers. One of the chief problems caused by this liaison is the overprescription of antibiotics, even for viral infections like colds (antibiotics don't work on viruses). If we aren't getting our antibiotics from doctors, we're getting them in our meat and milk. Of course, there's nothing wrong with the use of antibiotics if a person or animal comes down with a bacterial infection. Thank goodness we have them. The problem is their routine use on farms to prevent animals from becoming sick because they live in filthy, cramped conditions. Chickens crammed three to a cage in egg factories. Dairy cows slogging though their own manure in confined barns. Feedlot meat animals held in crowded pens. Pigs grown in huge warehouses so filthy you can smell them from miles away. All these animals are routinely given antibiotics so they don't fall prey to illness. It's said that nothing can live for long in its own waste products—unless it is kept on routine antibiotics. The result is that natural selection has created superbugs that resist most antibiotics, and these have become a huge problem in hospitals, nursing homes, and other institutions. Is the same true regarding the use of pesticides on conventional farms? You bet. Despite the applications of billions of tons of pesticides on American farm fields over the years, pesticide use continues because the resistant insects survive to breed. In some respects, though, American agriculture is changing faster than conventional medicine. The use of broad-spectrum pesticides (that kill whatever insects they contact) is shrinking in favor of less toxic compounds and more natural methods.

The answer is to eat and ferment organic foods and simply avoid the antibiotic problem altogether. If, for instance, an organic farmer has a cow with mastitis—an inflammation of the udder—that cow is removed from the organic herd, treated for its acute condition with antibiotics, and after the course of treatment has cured the cow and after a waiting period that allows the cow to eliminate the antibiotic from its system, the cow rejoins the organic herd. Antibiotics are a godsend, but only when used properly.

No discussion of the value of fermented foods for health is complete without an understanding of what probiotic science is just now revealing.

The intestinal flora defends the body against disease, primarily by colonization of the intestines by beneficial bacteria, thus preventing its colonization by pathogens. They do this by competing successfully with pathogens for nutrients, taking up the available attachment sites on the gut wall, producing bacteriocins, changing the pH of the intestinal contents to make it more acidic and less favorable for pathogens, and stimulating both innate and acquired immune functions.

While living lactobacilli are the most effective component of the flora for enhancing the immune system, yeast cell wall extracts that emerge as the cells decompose in the intestines encourage the body's production of immune system cells. Additionally, cells that make up the membrane that lines your intestinal wall produce mucus that speeds digesting food along its merry way, communicate with your body's immune system, have a two-way conversation with your brain, and perhaps most important, prevent the intestinal contents from penetrating the gut wall and entering your bloodstream.

There is a digestive disorder some call "leaky gut syndrome," and it's another one of those contested medical issues. Alternative health practitioners attribute many illnesses to the syndrome, but conventional medicine denies its existence as a valid diagnosis. The hypothesis holds that poor diet, an off-kilter intestinal flora, allergies to wheat gluten, parasites, and other dysfunctions create gaps between the cells that line the gut, allowing undigested food, feces, bacteria, and metabolic wastes to enter the bloodstream, where they can be carried to the farthest parts of the body.

The body's immune system springs into action and attacks these foreign particles, which can set off an autoimmune response that leads to diseases like lupus, Lou Gehrig's disease, and rheumatoid arthritis. Low-level inflammation from the leaking gut leads, in the opinion of those who believe in the leaky gut syndrome, to inflammatory bowel disease, heart disease, and plaque buildup in the arteries, as well as a range of other illnesses, including autism. While the leaky gut syndrome is not accepted by conventional medicine, it is at least a hypothesis that can be tested. According to the National Institutes of Health and the U.S. National Library of Medicine, conventional medical knowledge has no answer to the cause of autoimmune diseases, of which there are eighty types where the body's immune system fails to distinguish between healthy tissues and antigens that cause disease. What causes the immune system to no longer tell the difference between healthy body tissues and antigens is unknown, the NIH says.

Once again, it may just be fermented foods and our cherished friends, the live bacteria and yeasts that perform the fermentations, that come to the rescue. As Dr.

Loren Cordain, a member of the faculty of the Department of Health and Exercise Science at Colorado State University, and a leading expert on the Paleolithic diet, says, "When we have a healthy flora of bacteria in our gut, it tends to prevent leaky gut."

It stands to reason that probiotic foods like kefir, sauerkraut, kombucha, and the rest not only stimulate the growth of proper microorganisms in the gut, some of these microbes colonize the gut, healing it and plugging the gaps in the intestinal wall until the body can make the necessary repairs. Scientists are now studying substances that encourage the growth of healthy intestinal ecologies in our gut. Prebiotics are fermentable but as-yet-undigested foods, usually starches like whole grains, that increase the bifidobacteria and lactobacilli as they ferment.

So, given all the benefits of an intestinal tract heavily colonized by microbes, what does that say about those so-called colon cleanses?

You hear people talk about colon cleanses—how the colon is supposed to be plastered with old fecal matter and needs to be washed clean of accumulated toxins to support glowing good health. The idea has been around for centuries.

But the American Medical Association proclaimed the procedure invalid a hundred years ago. And with what we're learning about the colon these days, it seems sure that colon cleanses do more harm than good.

A review of twenty studies published over the past ten years shows that far from improving health and promoting weight loss, it's actually associated with bloating, vomiting, cramping, kidney failure, and even death. The practice, which involves laxatives, herbal teas, and pumping water into the colon through the rectum and evacuating it back out, can harm delicate colon tissue. And as we're learning in this book, our intestinal bacteria line the colon and communicate through the mucous lining with cells that function as part of the immune system. Why would we want to scour the colon of its natural layers of health-promoting bacteria?

About the only valid reason to clean the colon is to prepare it for a colonoscopy or radiological exam. And then it's best followed by a tall drink of kefir.

The portion of the public that ferments its food is similar to the organic movement of several decades ago. It's in the hands of aficionados now, sort of a

FERMENTED FOOD CAN PROTECT
THIRD-WORLD CHILDREN

Children in developing countries, especially in Africa, can contract many diseases caused by pathogens that infect the gut. Scientists at the Tanzania Food and Nutrition Center wanted to find out if food fermented with lactobacilli could help reduce the number and kinds of diarrhea-causing pathogens that infect children in Majohe village. They examined a group of 151 seemingly healthy children.

They found campylobacter, two strains of nasty *E. coli*, salmonella, and shigella pathogens in the children's feces. They then separated the children into two groups. One group was fed daily with togwa (cereal gruel) that had been fermented by lactobacilli. The second group was fed unfermented gruel of the same kind, also once a day.

All the children were tested for pathogens the day before the feeding trials started to provide a baseline. Rectal swabs were taken on day seven and on the last day of the feeding trials, day thirteen, and cultured to see the types and amounts of pathogens present in the children. The scientists waited another fourteen days, then took another swab and cultured it, to see if any benefit from the fermented food persisted. Here's the result:

Percentage of Children with Intestinal Pathogens

	Day 0	Day 7	Day 13	14 days later
Children fed unfermented gruel	27.6	11.4	8.1	22.6
Children fed fermented gruel	27.6	7.8	8.2	12.7

The scientists concluded that the togwa consumed once a day for three days a week helped to control intestinal colonization with potential diarrhea-causing pathogens in young children.

As you can see, while it took a couple of weeks for the children's intestines to be colonized by enough beneficial flora to curb the pathogens, the presence of infectious pathogens was cut nearly in half two weeks after the feeding trials stopped.

countercultural thing, but it's growing. More and more people are discovering how easy, healthful, and delicious it is to allow microorganisms to predigest some of their food.

People are beginning to wake up to the fact that the microbial world is at its strongest and most health-supporting when it is most diverse, and that strong, stable ecosystems of microbes are a major source of our own personal health around, on, and in our bodies. This is exactly what happened years ago when people started waking up to the fact that strong, stable ecosystems in garden and farm soils produced healthy, organic foods, and that those health benefits are transferable to the people who eat them.

It seems inevitable that in the future, more and more probiotic foods will become available commercially as companies rise to fill this niche. It's vitally important that these companies follow the lead of the organic movement in promulgating rules for the production of these foods—no chemicals, no shortcuts, respect for the microbes that are our partners in health—and insistence on wholesome ways these foods must be produced and handled.

CHAPTER 3

Probiotics and Genetic Engineering

There is a fly in the ointment of probiotics as human and animal health enhancers. Genetic engineers are working hard right now to modify probiotic microorganisms to manipulate their functions. But attempts to isolate the health-enhancing parts of whole organisms and transfer them to other creatures have been and will continue to be fraught with danger. Remember *Bacillus thuringiensis* (Bt), the bacteria that produced a toxin deadly to caterpillars (and caterpillars only)? It was a potent means of controlling pests safely that was used by organic gardeners and farmers for years. Genetic engineers found a way to insert the gene for production of the Bt toxin into the DNA of corn plants. The result? Instead of being used as a spot remedy, it became a wholesale killer of any caterpillars that happened into the corn or ate the pollen that drifted into neighboring fields—including the larvae of monarch butterflies. Whenever a wholesale onslaught is made against one of her creatures, especially microbes and insects, nature rushes adaptations to reduce the threat. As long as Bt was used as a spot remedy for a small area of caterpillar damage, there was no strong pressure for caterpillars to adapt and resist the toxin. But now that the genetic engineers have implanted it into corn's DNA, and there are 80 million acres of genetically engineered corn in the United States, there's enormous pressure for destructive insects like corn rootworm and corn earworm to adapt. And adapt they have. New scientific studies report that insects that formerly died when exposed to the Bt toxin have now developed resistance to it.

You may expect similar results when any organism—a whole system—is torn to its DNA fragments and then reassembled as components of other organisms. Such monstrosities were never planned by nature. Much genetic engineering is now being done on E. coli, a pathogenic intestinal organism. Should some unforeseen and very deadly Frankenstein version of E. coli escape the engineers' laboratory and head out into the public, we will discover how hard it is to track down every bacterium loose in a general population.

For example, Harvard Medical School researchers recently found many more genes that allow for antibiotic resistance in our indigenous gut bacteria than had previously been known to exist. They identified more than a hundred new genes that confer resistance to thirteen of the most common antibiotics. These new findings, along with previous tests, have now identified 215 genes conferring antibiotic resistance on gut bacteria. So of course the first thing the genetic engineers did was to stick these genes into E. coli to see if they transferred their resistance to the pathogen. And they did, at least in the lab. Could that transference to pathogens work in the gut? The study's results suggest that the answer to this question is yes. In other words, our overreliance on antibiotics in meat and dairy production and in human clinical treatments could have the perverse result of breeding antibiotic-resistant pathogens within our intestines. Antibiotics are a godsend, of course, but doctors around the world have been warning that their overuse is breeding super-bugs against which our current antibiotics are no longer effective. The new wrinkle is that it may be happening within our bodies.

In a rare instance of the demonstration of the transfer of genes from bacteria outside the gut to bacteria inside the gut in connection with food, scientists have found that bacteria in the guts of some Japanese people can digest porphyran, a polysaccharide compound in seaweed that is normally indigestible. This was discovered after a team of French scientists found the genes for a pair of porphyran-digesting enzymes in *Zobellia galactanivorans*, a marine bacterium. They then started looking for these genes in other species and found them in gut bacteria of Japanese folks who ate a lot of seaweed, but not in the gut bacteria of Westerners. They theorized that the human gut bacteria probably acquired the genes centuries ago in a gene transfer from marine microbes hitchhiking through the human gut on the seaweed so abundant in the Japanese diet.

Similarly, Westerners who eat land plants that contain polysaccharides can break down these compounds because of enzymes in their gut bacteria. "When you eat a salad, it's not you that breaks down the vegetables, it's the bacteria in your gut," said Gurvan Michel, one of the French scientists who found the seaweed-digesting bacteria. And maybe our gut bacteria acquired this ability from the microbes that had been breaking down land plants into their constituent nutrients for the past billion years: the soil bacteria that coat the plants we eat.

The intestinal flora as a whole consists of about 90 percent beneficial bacteria and 10 percent pathogenic organisms—not enough pathogens to overcome the advantage of beneficials in nutrient competition, gut wall attachment site availability, or the beneficials' production of compounds like bacteriocins that keep the pathogens disarmed or reduce their numbers. It's the same situation in the organic garden or farm, or in any healthy wild ecosystem, where there are always organisms with opposite functions that naturally form a healthy balance. In the microbial world, there are beneficial gut bacteria and pathogenic disease-causing bacteria, in the insect sphere there are plant-eating bugs and bug-eating bugs, and in the larger world there are chickens that eat grain and foxes that eat chickens. However, the range of illnesses caused by gluten intolerance—from full-blown celiac disease to simple gluten sensitivity—seems to upset this delicate balance of competing microbes in our gut, giving the pathogens more advantages and leading to conditions as uncomfortable as bloating to as life-threatening as colon cancer. While the intestinal flora doesn't cause gluten intolerance as far as we know, the health of the intestinal ecosystem can be thrown off balance by the condition. Such imbalance, which is called dysbiosis, is, at least in current medical thinking, caused by faults in the gut's secretions of peptide enzymes.

While we are focusing on these effects in humans, be aware that gut bacteria have all sorts of functions in the guts of other animals. For instance, it has long been thought that resistance to pesticides is always the result of an evolutionary leap encoded in the genome of certain insects. But according to a recent study reported in the *Proceedings of the National Academy of Sciences*, it's the presence of Burkholderia bacteria in the guts of *Riptortus pedestris*, a common bean bug, that is responsible. Burkholderia break down the insecticide fenitrothion, and 100 million of these soil-borne bacteria can inhabit the gut of a single bean bug. Bean bugs that harbored the bacteria survived doses of fenitrothion that killed 80 percent or more of undefended bean bugs within five days.

Within the intestinal tracts of all animals and within the living soil system that covers the earth's land are myriad processes that sustain life or digest life and that we are just beginning to understand. And such interactions don't just affect animals. It has recently been discovered by scientists at the University of Delaware that when disease organisms try to invade *Arabidopsis thaliana* plants through their stomata—those tiny pores on the undersides of leaves that open and close in response to environmental conditions—soil bacteria called *Bacillus subtilis* at the plants' roots signal the pores to close, shutting the door on the pathogens. The research involved inoculating three thousand arabidopsis plants with the leaf pathogen *Pseudomonas syringae*. The plants responded to the infection by recruiting *Bacillus subtilis* to bind to its roots, which caused the plants to manufacture abscisic and salicylic acids. The presence of these acids signaled the stomata to close.

The first complete human genome was sequenced in 2003, and since then scientists have been looking through our 22,000 genes for those involved in human disease. They have been successful to a degree, but some causes of genetic disease have proven elusive. The genes that cause the diseases don't seem to be in the human genome. That's where the Human Microbiome Project comes in—a worldwide effort by scientists to map the microbes that inhabit our bodies, starting with the 100 trillion bacteria in our gut. As we'll see, the cause of some genetic diseases may not be found in the human genome, because the genes aren't in the human genome. It may be that they are in the microbiome, that coating of the human body, inside and out, made up of microbes. While our human DNA has 22,000 genes, the DNA among all the kinds of microbes in our gut bacteria has 8 million.

The only human feeding study ever conducted on genetically modified foods shows that a foreign gene inserted into the DNA of soybeans spontaneously transferred out of the beans and into the DNA of gut bacteria, according to Jeffrey M. Smith in his book *Seeds of Deception*. The foreign gene produces a pesticidal toxin. The gene for production of the toxin, now residing in gut bacteria, allowed the toxin to contaminate the blood of thirty Quebec expectant mothers who ate GMO soybeans and then cross the placental barrier to contaminate the blood of their developing fetuses. The study appeared in *Reproductive Toxicology* in 2011.

If it seems incomprehensible that no long-term feeding studies in humans have been done concerning GMOs (genetically modified organisms), consider that to do so would require studying the effects of GMO foods on persons over many years, starting when they were children. There would have to be two groups of kids, one fed a diet of GMO food and another, a control group, fed exactly the same diet of non-GMO food. And there would have to be some force-feeding. If, for instance, the children eating GMO foods just picked at their vegetables while the non-GMO group readily ate them, then the GMO group would have to be force-fed the same amount of vegetables. Also, the groups would have to be isolated from the toxic chemical loads we all are exposed to, meaning they'd have to be away from home for many years. It stands to reason that it would be scientifically irresponsible and ethically immoral to conduct such a study. Better that we turn to animal studies, where we find compelling evidence from a 2012 French study that lab animals fed GMO crops for several years developed more and larger tumors, developed more kidney failure, and died earlier than a control group.

But to return to the study in *Reproductive Toxicology* showing the ability of the foreign gene for pesticide expression to contaminate the blood of mothers and their developing babies, consider that there are several serious implications. First, it means that the bacteria inside our intestines, newly equipped with this foreign gene, may create the novel protein inside of us. If it is allergenic or toxic, it may affect us for the long term, even if we give up eating genetically modified soy, because our contaminated gut bacteria will continue to express it.

According to a 2012 study published in the *Journal of Applied Toxicology*, low doses of the Bt toxin alone or in the presence of glyphosate herbicide (Roundup) kill human kidney cells. The study found that the Bt pesticide caused kidney cell death at concentrations of 100 parts per million, while Roundup at just 57.2 parts per million—two hundred times below agricultural use—killed half the test cells. "This study suggests that Bt toxins are not inert on human cells, and may indeed be toxic . . . Bt crops have previously been shown to induce liver and kidney abnormalities . . . in lab animals as well as immune responses that may be responsible for allergies," according to the study.

In response to studies like these, the French government has announced that they will not allow Bt sweet corn to be planted in their country—for reasons of environmental safety.

Though GMO crops are still allowed to be sold unlabeled in the United States, there is a simple answer for those of us who don't want weird Frankenfoods in our diet—eat organic. By law, organic foods must be GMO-free.

CHAPTER 4

The Conglomerate Superorganism

Scientists have begun searching the human body for the microorganisms that live on us and within us, and they've found them everywhere, not just as colonies of various microbes, but as whole ecologies of diverse microbes that have a profound effect on our body's development, health, and even behavior.

And they're everywhere on us. These ecosystems are in our mouths—some on the tongue, some on the lining of the cheeks, some on teeth, even with different ecosystems on different sides of the teeth. Some of our intestinal flora—the lactobacilli—are part of healthy mouth ecosystems that prevent tooth decay, while some are part of systems whose production of acid eats into enamel and encourages tooth decay.

There's an ecosystem up your nose, in your vagina, on your eyeball, and different ecosystems on the back of the knee and the front of the knee, on the wrist and on the back of the hand, and these differ from the ones on each digit of your fingers. If you compare the bacteria on two people's hands, only about 13 percent of the total will be the same. If you compare bacteria on one person's left and right hands, only about 17 percent of the bacteria will be the same.

You get the idea: We are not just ourselves, we are a conglomerate superorganism, a veritable landscape of microorganisms. Think of the landscape of a typical long-abandoned field in the mid-Atlantic states. In the open field are many types of grasses, each one adapted to its place in the system, some with intertwining roots, others with rhizomes that run spearlike through the soil. There are annual weeds and flowering plants and perennials that return from their roots each spring. Small trees and shrubs are there, too, and fruiting brambles, and wild grapes that hoist themselves into the taller trees. One could count a thousand different plants in any square mile and a hundred different ecosystems of plants, some adapted to low, wet spots, others that thrive on the rocky hilltops, and still others that are fit for the changes in light, climate, and water that characterize any rolling hillside and valley in that part of the country.

When it comes to microorganisms, your whole body is like that landscape. There are dark and light places, moist places, dry places, and all have the unique combination of microbes that find the situation suited to their needs. As birds of a feather flock together, different microbes that share an affinity for the differing places on your body grow there and form an ecosystem.

Scientists are just now beginning to catalog the various organisms that make up the human microbiome, as the whole panoply of microbes on us and in us is called. Some are obviously friendly helpers, such as the lactobacilli, some are obviously pathogens, like certain strains of E. coli, but for many, it's not entirely clear whether they are beneficial or harmful, or have both functions.

In a telling shift of focus representing the acknowledgment of large gaps in our understanding of the microbial life of our bodies, Julie Segre of the National Human Genome Research Institute in Bethesda, Maryland, says, "We've moved away from saying 'What are healthy bacteria?' to 'What are normal bacteria?'" She's been working on a catalog of skin bacteria for the National Institutes of Health's Human Microbiome Project. She cites acne as an example of a bacterial skin problem. "Is it healthy? I don't know. But it's normal," she told Science News.

The microbes in a normal intestinal ecosystem can change, depending on what the host human had for dinner and the host's wellness status, and those ecosystems can skew toward greater health or away from it. One study published in the journal *Nature* showed that the gut bacteria of people whose diets were rich in animal fats made more of a substance that leads to clogged arteries. On the other hand, a University of Maryland School of Medicine study found that healthy women had one of five kinds of bacterial ecosystems in their vaginas. Four of these were dominated by lactobacilli that made infection-preventing lactic acid, while the fifth ecosystem had few lactobacilli. You would think that this latter group of women might have an infection, but no, they were perfectly healthy. The bacteria in that ecosystem, though they weren't lactobacilli, also made lactic acid. So with our microbiome, it's good to look at what the microbes do, not just who they are.

For example, take the case of *Bacteroides fragilis,* which prevents and can cure inflammatory bowel diseases in animals, including humans. It does this by making a polysaccharide coating that it applies to the intestinal wall, and this coating prevents pathogens from getting a toehold to do their dirty work. How nice. Does *B. fragilis* love its human hosts so much that it plays doctor in the gut? Well, it turns out that it has an ulterior motive. The polysaccharide coating calms the host's immune system. Without it, the immune system would read the presence of *B. fragilis* as an infection and throw the bums out. So to keep itself healthy and happy in the human microbiome, it has learned to protect itself by protecting us. It also turns out that *B. fragilis* has been around for 500 million years, so as the human body developed through its entire evolutionary process, *B. fragilis* accompanied it, learning to do what it needed to survive in the pleasant climes of the human gut.

Does this mean that our microscopic friends actually manipulate us for their own ends? Yes, it does. Scientists in Sweden and Singapore, working together, found that mice raised in a sterile environment (without gut bacteria) were more likely to take risks than mice with normal gut flora. Normal mice tend to skulk along walls and avoid bright areas, but the sterile mice showed more willingness to walk freely in the center of their room and in brighter light. So somehow the presence of gut bacteria in mice leads them to be more wary, which increases their chance of survival, and by extension, the chance of survival of the gut bacteria that call the mouse home.

When very young sterile mice were given gut bacteria, their behavior reverted to normal skulking. But when adult sterile mice were given intestinal flora, their behavior didn't change, leading the scientists to theorize that these behavioral changes are locked into the brain at an early age. But mouse studies are one thing, and aren't always true for humans. Still, Sven Pettersson, a prominent Swedish microbiologist working on how gut bacteria signal the human brain, has stated that he wouldn't be surprised to find that bacterial signaling acts on pregnant mothers and affects the developing child.

It stands to reason that our modern world's constant attempts to control and eliminate microbes from our environment can have some pretty disastrous consequences. We have coevolved with a huge diversity of microbes, and in some measure, what and who we have become are predicated on their presence and actions. Take the example of *Helicobacter pylori*, a stomach bacteria that causes ulcers and even stomach cancer. The number of people who are constantly exposed to antibiotics in the meat and milk they ingest continues to increase, and the number of people carrying *H. pylori* in their stomachs continues to decrease. But, as with many members of our microbiome, bacterial actions aren't all bad. *H. pylori* produces an antacid substance that reduces the incidence of acid reflux, where stomach acid backs up into the esophagus and causes heartburn at best and esophageal cancer at worst. Also, people taking antibiotics to wipe out *H. pylori* have higher levels of a hunger-inducing hormone that may be contributing to the epidemic of obesity in our society. And it's been shown that people with *H. pylori* in their stomachs have lower risks of getting childhood asthma and allergies. The point is that *H. pylori* can be beneficial—or require antibiotics when it causes stomach ulcers. The key is finding out how to reduce its potential for causing illness and maximize its potential for health benefits. I suspect the key is in flooding the gastrointestinal system with a healthy, diverse mix of microorganisms through ingestion of fermented foods.

The work on *H. pylori* is focused on just one internal microbe, but antibiotics wreak havoc on whole swaths of our microbiome. One study showed that antibiotics altered the levels of 87 percent of the compounds made in mouse intestines by the animals' intestinal flora. Many of the biological functions that were negatively affected, including the production of bile salts and steroid hormones, are also important for human health.

It shouldn't be too surprising that the life forms with which we coevolved—big people and tiny microbes—are so closely interconnected. Healthy ecosystems are always in balance, including the ecosystems on and in our bodies. When we big people misuse antibiotics, sterilizing agents, antibacterial soap, and other chemicals that upset nature's balanced microbial colonies, disease can result. As Rob Knight, a microbial ecologist at the University of Colorado at Boulder, put it in an interview for *Science News*, "Antibiotics are like driving a bulldozer through your garden and hoping that what pops back up is what you want." My experience with organic gardening shows me that what pops back up is not only not what you want, but is more likely to be what you don't want: opportunistic rough weeds are always the first to recolonize disturbed soil. There are good reasons for this. The rough weeds, like giant ragweed, pigweed, and lamb's-quarter, are what's called C4 plants, while our crop plants, like carrots and spinach, are C3. Without getting too technical, the difference is that C3 plants respond to heat and stress—disturbed soils are generally hotter and drier—by closing their stomata, which prevents the plants from inhaling carbon dioxide or exhaling oxygen. Oxygen thus builds up within the C3 plants and shuts down photosynthesis, stopping their growth. C4 plants have adapted by using a different enzyme to catalyze carbon, one that reduces the buildup of oxygen and allows the plants to keep growing even in conditions of heat and drought. Now, nature not only abhors a vacuum, she abhors bare soil because sunlight, heat, and dryness can harm her precious soil microbes. So she gives the advantage to C4 plants as the first responders when soil is plowed up because they do a better and quicker job of shading the soil with their leaves. That's why you may plow down garden crops only to find that what pops back up is a lot of ragweed and pigweed.

Similarly, with our intestinal garden. There may well be times when a course of antibiotics is needed, but then the gut is vulnerable to pathogens. So it's important to keep the gut well supplied with lots of live-culture fermented foods every day during the course of the antibiotic treatment. Yes, most if not all of these microbes will be killed by the antibiotics, but because you are loading more in regularly, the intestines will still be coated with beneficial microbes that keep pathogens at bay.

Just as most landscapes on earth are ecosystems that have not been studied in scientific detail, so our microbiome is still just being looked at. "It's as if we have these other organs, and yet these are parts of our bodies we know nothing about," Dr. George Weinstock of Washington University in St. Louis told the *New York Times*. Dr. Weinstock is part of the Human Microbiome Project's international effort to catalog new species of microbes from our bodies, not by looking at them under a microscope—many of the microbes die off when being transported to the viewing table—but by gathering their DNA. They scraped the skin, swabbed the cheek, and sampled the surfaces at eighteen sites on the bodies of three hundred volunteers. What they get is a jumble of millions of DNA sequences left by hundreds of different species of microbes. How to make sense of it?

That's what the Human Microbiome Project is attempting to do. First researchers sequenced the entire genome of 900 species of microbes that can be cultivated in the laboratory. From these they discovered nearly 30,000 genes that are unlike any other known genes, which was surprising because these were microbes that could survive the trip to the viewing table and had been studied for years.

The new work is opening doors on a truly multifaceted world of microbial ecosystems in and on our bodies. Dr. David Relman of Stanford University, who's participating in the project, estimates there are between five hundred and one thousand species of microbes in our mouths alone. "It hasn't reached a plateau yet," he told Carl Zimmer of the *New York Times*. "The more people you look at, the more species you get." And the mouth's subdivided ecosystem species vary from person to person as well. Only about one hundred to two hundred species live in any given person's mouth at a given time, but all the rest are available for colonization, and indeed the composition of the mouth's ecosystems may change widely over time.

The project is demolishing some old ideas as well as highlighting new ones. For many years, the lungs were thought of as sterile tissue, but scientists have now found 128 species of bacteria on the lungs of healthy people. Every square centimeter of lung tissue, it turns out, carries about two thousand bacterial inhabitants. All the microbial permanent inhabitants, all the renters so to speak, and the hitchhikers, plus all the cells of the human body make up the human biome, and so the DNA of all the biome's creatures, including "us," is an enormous mishmash, and the total number of genes has to be mind-boggling.

Microbiologists from the University of Puerto Rico recently spent time studying the biome of newborns in a Venezuelan hospital. Those who were born vaginally were covered with microbes from the mother's birth canal, while babies born by Caesarean section were coated with microbes typically found on the skin of adults—two very different microbial ecosystems. The microbiologists thought that the C-section babies would be sterile, like they are in the womb, "but they're like magnets," said Dr. Maria Dominguez-Bello, who was surprised at how quickly the sterile newborns delivered by C-section became colonized.

We're also finding out that babies born by C-section may be more likely to become obese children than babies born vaginally. A study published in the *British Medical Journal's Archives of Disease in Childhood* found that babies born by C-section were about twice as likely to become obese as those delivered vaginally—15.7 percent compared to 7.5 percent. Even after accounting for the mother's weight, the length of time being breast-fed, and the baby's size, the numbers held up. The study's authors suggested that delivery by C-section alters the babies' acquisition of key digestive bacteria in their intestinal flora and that this might be the cause of the high levels of obesity. About a third of the births in the United States are now by C-section, compared to just 20 percent in 1996.

Children born by C-section are shown to be more likely to get infections from methicillin-resistant *Staphylococcus aureus* (MRSA) than children born vaginally, possibly because they lack the healthy gut bacteria seeded in their intestines by the mother's healthy vaginal ecosystem. C-sections have also been linked to an increase in asthma and allergies, as have antibiotics. Young bodies need signals from internal microbes to fully and properly develop. For instance, mice grown in a sterile environment with no gut bacteria develop stunted intestines.

Scientists are now discovering that certain diseases are accompanied by changes in the biome. Asthma sufferers harbor a different microbial ecosystem than healthy people, and obese people have different gut bacteria than people of normal weight. All this information leads toward the new understanding that when children fail to develop a strong, diverse ecosystem of gut bacteria, their immune systems aren't trained to fight diseases in the normal way. Their immune cells may too vigorously cause inflammation that damages the person's body instead of protecting it.

Besides causing trouble, changes in the biome, or at least the intestinal part of it, may support healthy, normal bodily functions, like gestation. Research led by Ruth Ley at Cornell University and published in the journal *Cell* shows that women's gut microbes become less normal and less diverse as pregnancy progresses. The researchers examined stool samples from ninety-one pregnant women in their first, second, and third trimesters and discovered that changes in the gut bacterial ecosystem that normally cause weight gain and inflammation may actually benefit expectant mothers. As the nine months passed, the pregnant women's gut bacteria became less normal and diverse. They also found the number of beneficial bacteria declined while the levels of disease-related bacteria increased. The changes were not related to diet.

"By the third trimester," Ley reported, "the microbiota can induce changes in metabolism." Ordinarily, these changes can lead to type 2 diabetes and other health problems, but in the context of pregnancy, they're beneficial because they promote energy storage in fat tissue and help support the fetus. The takeaway from Ley's study is that our bodies have coevolved with our intestinal flora and that the flora cause healthy changes in the body's metabolism while the body readies a developing fetus to term.

The colonization of our bodies proceeds throughout life, every day, because we live and move in a torrent of microbes, without which we couldn't survive. According to scientists taking part in the Human Microbiome Project, human beings on their own produce a paltry number of enzymes that break down plant matter into its constituent nutrients. The wide range of microbes in our guts, on the other hand, have a large arsenal of enzymes that perform this critical function, and they perform it on our behalf.

I can lose any number of body parts—a finger, an eye, a leg—and still be me and be healthy. But I can't lose my intestinal flora. Without it, my immune system is compromised, my ability to digest plants is compromised. I wouldn't last long. So what and who am I?

CHAPTER 5

New Findings About the Ecosystems Within Us

In nature, there are various ecosystems that we can see with our naked eyes, organized by the kinds of environmental conditions that exist in their habitats. From New England across the northern tier of states to the Pacific Northwest, you can find long-abandoned field ecosystems, with grasses and tall weeds growing in the sunlight, with old apple trees in decline, left over from a time when the field was an orchard, all abutting woodland. Where water ponds and flows, you will find a riparian ecosystem that's quite different from the old field, with almost completely different weeds and shrubs, water plants, and fauna. Forests and mountains in the Rockies and desert ecosystems in the Great Basin are all very different, with different players participating.

There are also ecosystems that we can't see, including those in our intestines. And if you examine people's gut ecosystems closely, as researchers at the European Molecular Biology Laboratory in Heidelberg, Germany, have done, you will find something astonishing. The scientists have discovered that people's gut ecosystems fall into three distinct types, which they call enterotypes. And the three categories hold up despite differences in ethnicity, sex, weight, height, health, or age. The scientific team is now exploring possible reasons why the three enterotypes exist. It's all speculation at this point, but it may be because the intestines of newborn infants are randomly colonized by certain collections of pioneering bacteria, and these microbes change the gut so that only certain species can follow or elaborate on them. We will surely find the answer in coming years.

As we know, gut bacteria help in food digestion, work with the immune system, help prevent disease-causing bacteria from gaining a toehold in the gut, and synthesize vitamins by producing enzymes that our bodies can't produce. Dr. Peer Bork, the leader of the Heidelberg group, and his fellow scientists have found that each of the three enterotypes makes a different balance of these enzymes, and thus produces a different mix of vitamins. For example, the ecosystem they call Enterotype 1 (ET1) has high levels of bacteroides that produce enzymes for manufacturing vitamin B_7, also known as biotin. Bacteroides in Enterotype 2 (ET2) were fairly rare, while the genus *Prevotella* was very common and produced enzymes that make vitamin B_1, called thiamine.

I found no mention of the predominant bacteria or their function in Enterotype 3, so I asked Dr. Bork about it. In an e-mail, he replied, "With more than six hundred samples analyzed, Methanobrevibacter seems the major driver of Enterotype 3, but also Ruminococcus is associated with ET3. Haem biosynthesis seems overrepresented in ET3." Rather than producing enzymes that allow the synthesis of vitamins, Enterotype 3 bacteria are evidently involved in heme ("heme" is the American spelling; "haem" is the European) biosynthesis—the production of hemoproteins, the most familiar of which to most people is hemoglobin. There are many other heme proteins with various important functions in the body. *Methanobrevibacter smithii* is the dominant archaeon (a class of microbes with no nucleus) in the human gut. It is important for the efficient digestion of complex sugars because it consumes end products of bacterial fermentation, helping to transform nutrients into calories. *Ruminococcus* is an anaerobic, gram-positive microbe. One or more species in this genus are found in significant numbers in the human gut.

Dr. Bork told Carl Zimmer of the *New York Times* that doctors might be able to use knowledge of the enterotypes to find alternatives to antibiotics. Instead of trying to kill disease-causing pathogens that have disrupted the ecological balance of the gut, Zimmer wrote, they could try to provide reinforcements for the good bacteria by attempting to restore the enterotype you had before. This provides probiotics with a new wrinkle.

Dr. Bork said his team had no hypothesis when they started comparing ecosystems of gut bacteria among people both healthy and sick, obese and slender, Asian and European, and so on—four hundred people in all. They just wanted to see what was happening, and they were startled to find that the gut ecosystems all fell neatly into the three categories. These findings concern gut bacterial ecosystems—broad, overall categories of microbes that have a dominant type of bacteria for each category and specific functions within our bodies. If you compared these gut ecosystems to life at the macro level where human beings live, it would be like saying beavers and trout dominate the streams and lakes ecosystem of Maine, while alpine wildflowers and eagles dominate the Rocky Mountains' higher elevations, and deer, bear, and hickory trees dominate central Pennsylvania. All these systems are healthy, they all share some of the same flora and fauna, but each is different from the others.

This is just the beginning of our understanding of the universe of microbes on us and in us. For every person on earth, there are almost 100 trillion cells on or in his or her body, only 10 trillion of which are human body cells.

From our understanding of how health develops in natural ecosystems, whether garden or body, we can see why the surest way to develop an unpleasant body odor is to use harsh deodorants and antiperspirants. By cleansing yourself of microbes with antibacterial soaps, you are dismantling their healthy ecosystem on your skin. The sterile skin breathes and exudes sweat and other natural compounds, and the first microbes to recolonize the skin start to metabolize the sweat and other compounds. The process develops as an unpleasant body odor. The answer is to let the skin's healthy ecosystem of microbes reestablish itself by using pure soap without antiseptic chemicals or detergents daily and putting away the harsh deodorants.

TRUST YOUR GUT

Most people don't think much about their intestines—their guts—unless they are speaking metaphorically.

Think of the way people talk about guts:

"They have guts," meaning they're brave, stand up for what they believe, show courage, do the right thing even though it's difficult. Some corollaries have become clichés: "No guts, no glory." "It's hard, but he's going to gut it out." And so on.

Other metaphorical uses describe intuition: "Trust your gut." "I had a gut feeling." "I just felt in my gut it was the right thing to do."

And those gut feelings are also where you measure something against reality. To make a "gut check" is to see how you really feel about something, not with your head, or even with your emotional heart, but with the integrated whole being of who you actually are.

In other words, the gut is where you confront your personal feelings, priorities, and desires, even more than with your head and your heart. And it's where you act from those feelings without, or despite, fear of the consequences. Could it be that the biological systems within us, without which we couldn't function as human beings, and on which we depend for life itself, are so integral to who we are that when we need to check in on who we really are and make crucial decisions—even life-and-death decisions—we take the gut into account?

Most of us would like to think of ourselves as just the hominid portion of our being and the intestinal flora as something else. But it may be that those two seemingly separate entities are one, and that we reveal that when we speak of our guts.

A recent *New York Times* editorial, titled "A Universe of Us," declared that "we think of ourselves as individuals . . . but whatever else we are, we are also a complex ecosystem, a habitat." Scientists have now discovered another realm within that habitat that they call the virome—a large community of viruses. Far from making us sick, these viruses are a part of the human biome that makes us healthy.

Microbiologist Jeffrey Gordon reported in a study published in *Nature* that each of us has a pattern of viral DNA that is highly stable and highly distinct, even among closely related humans. This differs from our bacterial communities, which tend to evolve over time and are similar among family members.

That means you and I are not just the expression of our individual human genome and the DNA in our human cells, but rather we are a genetic landscape with the DNA of hundreds of different species working together to keep us healthy so we have time for wondering what's for lunch.

CHAPTER 6

The Intelligent Intestine

It appears that the old saying "The way to a man's heart is through his stomach" is scientifically accurate as well as metaphorically true—if by heart you mean the place where his thoughts and desires originate; that is, his brain.

The human gut communicates with the brain, recent studies show. Now there's evidence that it's the gut microbes that communicate directly with the brain. A healthy ecosystem of friendly gut bacteria not only keeps the gut happy, it evidently may help keep their hosts happy, too, *Science News* reports.

Certain parts of the brains of mice fed a nutritious broth containing *Lactobacillus rhamnosus* developed more of a protein that senses an important chemical messenger called GABA than mice fed a sterile broth. The result was that the mice fed the gut bacteria were less anxious than mice fed sterile broth. In stress tests, the probiotic mice had fewer stress hormones than the sterile-fed mice. Further research showed that the chemical messengers leading to reduced feelings of stress and anxiety most likely traveled from the gut to the brain via the vagus nerve. In humans as well as mice, the vagus nerve connects the brain to the viscera, regulating gut function and reporting the state of the gut to the brain.

While mouse experiments can't necessarily be extrapolated to humans, these studies do open up potentially important areas for human study. "Mouse experiments hint that gut bacteria could play a role in a wide variety of brain and psychiatric disorders such as depression, autism, and schizophrenia," the *Science News* report said. In its conclusion, the scientists reported that the effect on the brain and subsequent behavioral effects were not found in mice whose vagus nerve was cut or removed, thus identifying the vagus as a major communication pathway between the bacteria in the gut and the brain. The study concludes that gut bacteria play an important role in communication between the gut and the brain axis. It may be that certain intestinal organisms can be useful in developing therapies for stress-related disorders such as anxiety and depression.

As you can probably attest, delivery of a delicious, satisfying meal to the stomach and the alimentary tract beyond can induce feelings of well-being and satisfaction. That very well could be your intestinal flora sending a thank-you note to your brain.

Bacteria inside the intestine also communicate with the mucus-producing cells that line the organ and with the body's immune system that is working at the intestinal site, and all of this is to protect the host (that's you and me). "The intestine is the primary immune organ of the body," according to a report published in the *American Journal of Clinical Nutrition,* a peer-reviewed scientific journal.

It stands to reason when you think about it. Into the mouth goes food that's covered with mold spores, bacteria of various types, fungal spores, and other microbes in numbers that reach the trillions. And this is good fresh food. Among the food's payload of microbes are some pathogens that can cause sickness and disease. But if we're careful, very few times will our food's bad guys cause illness. And that's because the passage of food through the digestive system is a gauntlet of peril for pathogens. The gut is lined with elaborate systems for the prevention of illness because from the time our distant ancestors started eating, mixtures of microbes have been passing through evolving guts that have been refining and improving their ability to ward off illness.

The colon's permanent immune system components—colonies of beneficial bacteria, the mucosal barrier, and the local immune system—literally communicate with one another, and with the temporary, in-transit bacteria provided by certain probiotic foods such as yogurt and sauerkraut.

The Gut and Psychology Syndrome

The link between intestinal flora and the brain has another champion in Dr. Natasha Campbell-McBride, a Russian neurologist and neurosurgeon who has a postgraduate degree in human nutrition and has been practicing for two decades in Cambridge, England. She attributes a dysfunctional intestinal ecosystem as a root cause for illnesses as disparate as ADHD, ADD, dyslexia, depression, schizophrenia, and, most intriguingly, autism. This is a controversial area, and conventional medical practitioners and scientists have cast doubt on the role of a compromised gut in these conditions. I don't know if Dr. Campbell-McBride's idea is correct, partially correct, or off base, but there is enough evidence to suspect that she's on to something fundamentally important in medicine, and that is the role of a healthy intestinal flora in human well-being.

Dr. Campbell-McBride has some firsthand knowledge of a possible link between a dysfunctional gut and autism. Her firstborn son was diagnosed with autism at age three and she soon realized that her specialty of neurology had little advice for how to treat him. She also noted a rapid rise in autism, from one child in 10,000 in 1984 to one child in 150 by 2005, and today the number is one in 66 in the UK and one in 88 in the United States, with similar numbers in Australia and New Zealand. Her studies in human nutrition led her to discoveries about intestinal flora, which she applied to her son, changing his diet to strengthen his gut ecology. Today he is no longer autistic.

Dr. Campbell-McBride believes that autistic children have perfectly normal brains and sensory organs when in the sterile environment of the womb. The trouble starts when their mothers reach term and they enter the birth canal. "What happens in these children [is that] they do not develop normal gut flora from birth," she says. Instead of a healthy intestinal flora picked up from a mother with a healthy gut ecosystem, the child's intestines are colonized by pathogens. She believes that as a result, the child's digestive system becomes a major source of toxicity, with pathogenic microbes damaging the gut wall, allowing toxins and microbes to flood into the bloodstream and reach the child's brain. These toxins, she posits, interfere with the ability of the child to process sensory information. They turn the sensory information into noise, from which the child can't learn how to communicate, how to understand and use language, or how to develop all the natural behaviors and coping behaviors that normal children develop. She calls this abnormal condition Gut and Psychology Syndrome, and it is the cause of autism in her opinion.

It should be recognized that Dr. Campbell-McBride's opinion is very controversial, and often debunked by the scientific community. However, current research into autism shows that one of its root causes may be chronic inflammatory disease, which scientists call immune dysregulation, that can occur in a pregnant woman. This includes allergies and autoimmune diseases like rheumatoid arthritis. A large Danish study that tracked 700,000 births over a decade found that a mother's rheumatoid arthritis raised a child's risk of autism by 80 percent and that if the mom had celiac disease, an inflammatory disease caused by proteins in wheat, while pregnant, the child's risk of developing autism was 350 times greater than children of mothers who had no inflammatory disease during pregnancy. This study suggests that the mother's immune system is constantly flooding her body—and the fetus's—with pro-inflammatory molecules. Cells in the baby's developing brain that help build and maintain neurons are enlarged by chronic activation. The brain's wiring gets scrambled. That's the supposition.

So, if many cases of autism are related to immune dysfunction, what has happened to our modern immune systems that has caused such a precipitous rise in its prevalence? Writing in the *New York Times*, Moises Velasquez-Manoff, author of *An Epidemic of Absence: A New Way of Understanding Allergies and Autoimmune Diseases,* says that "scientists have repeatedly observed that people living in environments that resemble our evolutionary past, full of microbes and parasites, don't suffer from inflammatory diseases as frequently as we do."

The bottom line of this train of thought is that if inflammatory disease can be reduced during pregnancy, there's a chance that autism rates could be reduced, too. There hasn't been enough science done on this topic to know whether this hypothesis will hold water when it's tested. But there are so many good reasons to eat fermented foods that it's certainly wise for pregnant women and women planning a pregnancy to keep their intestinal ecosystems strong and healthy.

Dr. Campbell-McBride says that the baby acquires its intestinal flora from the mother's vagina at birth, so whatever microbes live in the vagina become the baby's flora. The vaginal flora comes from the bowel, so if the mother has abnormal gut flora, there will be abnormal flora in the birth canal. She thinks that the epidemic of abnormalities in gut flora began during World War II, when antibiotics were discovered. Antibiotics tend to wipe out the beneficial bacteria, and that offers a window of opportunity for the pathogens to proliferate. Given the routine use of antibiotics in the production of meat and milk in the countries with high autism rates, one can theorize that these antibiotic residues are constantly harming the diversity and extent of the intestinal flora.

"What I see in the families of autistic children is that 100 percent of the moms of autistic children have abnormal gut flora and health problems related to that," Campbell-McBride says. Again, this is merely anecdotal evidence and not hard science, but it suggests a possibly fruitful area for science to investigate and it makes sense for moms-to-be to ingest probiotic foods that promote a normal and healthy intestinal flora.

Once more, there's the parallel with what we see in agriculture and horticulture. If we wipe out the so-called pest insects, we are really destroying the whole insect ecosystem, because the beneficial insects are more susceptible to pesticides than the pests, and the so-called pests are food for the beneficials. So the farm or garden is effectively sterilized of insects. The first insects back in won't be the beneficial insect-eating bugs because there will be no food for them. It's the pests that return first, because now the field lays open to them, with no predation from the good guys. And once again, the same kind of ecological pressures affect the world of microbes, with the same results: a proliferation of pathogens after a sterilization.

It's vitally important for women of childbearing age to establish a healthy, strong intestinal flora. The easiest way to do that is to first eat organic foods that by law do not contain pesticides, antibiotics, or genetic modifications; second, seed their intestines with a diverse mix of microbes by regular ingestion of probiotic foods such as kefir, yogurt, kombucha, sauerkraut, and the rest of the healthy fermented food family; and third, breast-feed the baby, which bathes the child's intestinal tract with food for healthy gut bacteria that nature will provide in abundance.

If for no other reason, minding the health of your intestinal flora makes you a healthy mom—and dad, too, for dad's flora colonizes the groin and is shared with mom during intercourse. And a healthy mom and dad can only be beneficial for baby.

There's a further thought that this information brings up. It seems that our intestinal flora—nine out of every ten cells and 99 percent of the DNA in our bodies—really does communicate with our brains, affecting our moods as well as our physical and mental health. An intestinal tract damaged by chemicals and rife with shattered ecosystems and pathogenic organisms talks to the brain in a broken and misinformed voice, so to speak.

But a healthy, diverse ecosystem of beneficial microbes sequesters pathogens, prevents them from getting a toehold in the gut, destroys them with bacteriocins, and in other ways keeps them from harming us. It eliminates the static that shreds meaning. It speaks clearly and plainly.

It's possible that the healthy communication between gut and brain has much more to do with who we are than we allow. We think of ourselves as autonomous beings with free choice and the ability to screen out unwanted information from our environment. But we can't screen out the communication that our flora and brain share out of our conscious sight, deep within the silence of autonomous ner-

FERMENTED FOODS FOR A SEXIER YOU

Male mice fed bacteria-rich yogurt developed a certain swagger to their gait, according to an unpublished study by cancer biologist Dr. Susan Erdman and evolutionary geneticist Eric Alm, both at the Massachusetts Institute of Technology. This was caused by the males projecting their testicles outward. And this happened because the testicles were 5 percent heavier than those of mice fed a typical diet and 15 percent heavier than mice fed junk food.

As a result, the males fed yogurt inseminated females faster and had more offspring. Females fed yogurt had larger litters and were more successful in raising their pups to maturity. "The probiotic microbes in the yogurt help to make the animals leaner and healthier, which indirectly improves sexual machismo," the researchers reported. Now scientists at Harvard are investigating the effect of yogurt consumption on semen quality in human males. So far, they say, their results are consistent with those on mice.

vous systems within us. And this communication doesn't stop when we sleep, because bacteria and other microbes don't sleep. That communication with the brain continues. Scientists know that the brain continues to function during sleep—and we know that, too, because we dream, and because our unconscious eyes jitter as they search the emptiness during REM sleep. Could mood-improving intestinal microbes keep nightmares away? If they can improve our mood during the day, why not at night? After all, we all remember Ebenezer Scrooge's reaction when confronted by Marley's ghost in *A Christmas Carol*. Scrooge at first doesn't believe in the ghost, and the ghost says to him, "Why do you doubt your senses?"

"Because," says Scrooge, "a little thing affects them. A slight disorder of the stomach makes them cheats. You may be an undigested bit of beef, a blot of mustard, a crumb of cheese, a fragment of an underdone potato. There's more of gravy than of grave about you, whatever you are!" Ever prescient, Dickens knew that bad food could lead to bad dreams. So did the great cartoonist Winsor McCay, who created *Dreams of the Rarebit Fiend* in the early twentieth century (he also did *Little Nemo in Slumberland*). In these fantastically well-drawn strips, the protagonist each time finishes a rarebit and sinks into a haunted nightmare. Perhaps a few forkfuls of sauerkraut or a long draft of kefir would have prevented that.

The gut flora looks out for us, for we are its hosts and carriers. Without us, the flora is lost. And so a healthy flora whispers its healthy signals to the brain. Conversely, without our healthy gut flora, we are lost. It's time for humanity to realize what's really going on within us and to cherish our symbiotic flora by giving it what it needs to grow strong and healthy.

It's literally true that our body and our flora are functionally one being.

A bacterium called *Mycobacterium vaccae* acts as an antidepressant when it gets into your bloodstream. It causes people to produce more serotonin and norepinephrine, two compounds that make people feel happier. The bacteria also boost the immune system, vitality, and general cognitive function, according to a paper published in the journal *Neuroscience*.

The bacterium was originally isolated from the feces of cows (hence *vaccae*), meaning it's an inhabitant of the bovine gut. It's also found in good organically treated soil. We may be inhaling the bacteria when we're working with compost and digging in our gardens, if not ingesting it from our food. But it appears that both *Lactobacillus rhamnosus* and *Mycobacterium vaccae* stimulate the production of neurologically active brain chemicals that elevate our moods and increase our enjoyment of life.

The effect of the bacterium on human mood was discovered accidentally about a decade ago by Dr. Mary O'Brien, an oncologist at the Royal Marsden Hospital in London. She injected patients with killed *M. vaccae*, expecting that the bacteria would stimulate the immune systems of lung cancer patients. It did that, but also improved her patients' emotional health, energy, and mental function.

Intrigued by her observations, Dr. Chris Lowry and colleagues at the University of Bristol followed up with a similar study of this effect and reported their findings in an issue of *Neuroscience*. They achieved similar results.

CHAPTER 7

A Day Spent in Berkeley with the Fermenters

The city of Berkeley, California, is a progressive place where the art and science of fermentation have already sunk deep roots, spawning businesses that provide cultured foods and beverages as well as services.

For instance, I heard about a marvelous lacto-fermented squash dish created by Angie Needels, director of MamaKai in Berkeley, and wanted to taste it. Her business is to deliver ready-made meals to pregnant women, women nearing term, and new and lactating mothers in the postpartum stage.

On a sunny morning in February, I pulled up at Ms. Needels's neat frame house on a quiet Berkeley street, hoping to get a recipe or two. What I got instead was a glimpse into the real value that adding fermented foods and establishing a healthy gut ecosystem represent for families in their childbearing years. I've raised five kids, and I know from personal experience what a load of work it can be, especially for Mom, even if Dad helps share the load; both parents can be exhausted during the process.

Angie, a very bright and energetic young woman, studied fine art at the Columbus School of Art & Design in Ohio, and worked in the insurance field there before she and her husband moved to the Bay Area to inject some change into their lives. Once settled in Berkeley, she discovered the region's wonderful abundance of all kinds of fresh and organic foods and decided to explore a culinary career by enrolling at Bauman College, which teaches holistic nutrition and health. "One of my class assignments was to develop a business plan involving whole, nutritious foods," she said. While a lot of her classmates were focusing on what kind of restaurant they might develop or how they could help folks heal and cure their illnesses, Angie recalled how often her pregnant friends or those who had just given birth said they'd love someone to come and cook for them.

"I thought to myself, 'What if I start before people get sick or exhausted,'" she said, "and work in the time from preconception to preschool to help women toward really good health?" She saw that the postpartum period was difficult for young women used to being socially active. Now with the work of caring for baby, making the meals, and doing housework, there wasn't much time in the day for some "me" time, and young mothers can feel isolated. So she conceived the idea of a service to provide optimal nutrition for pregnant women and new mothers. In choosing a name for this business, she knew that *kai* is the Maori word for food, and so she named it MamaKai, meaning mother-food.

"I don't thrive on dairy, but I do thrive on fermented foods, and my food for moms reflects that," she said. "From the time the baby is born and for the first three or four weeks of life, the baby's digestive system is just being set up. The first intestinal flora comes from the birth canal, and the flora in the birth canal reflects the flora in the intestine. Fermented foods establish a healthy intestinal flora, and that's transferred to the baby. But there are certain foods like dairy and cruciferous (cabbage family) vegetables, for instance, that affect the breast milk." As she said this, I thought of my days as a teenager growing up next to a dairy farm, and how in the spring the cow's milk would smell and taste of the wild onions and annual mustards they'd graze. "I encourage moms to avoid eating dairy and crucifers for two to three months, until the baby's digestive system gets more mature. I also suggest

reducing the amount of glutinous foods, like grains, breads, and beans."

The result, she said, is almost miraculous. Babies have no colicky discomfort, no digestive upsets, no crying jags because of gas. And they sleep better, longer, and more soundly, often right through the night. It really makes for happy, healthy babies and reduces anxiety and sleepless nights for the parents. "It's only for two or three months, then moms can go back to eating what they like, although I recommend continuing with a diet rich in fermented, probiotic foods."

During this time when MamaKai is delivering food for the new mom, Angie swears by what she calls bone broths. The bones of beef, pork, lamb, and goat go into a big stockpot, along with onions, carrots, celery, carrot greens, and seaweeds, and the pot is simmered for thirty-two to thirty-six hours, until the liquid is reduced by half. Bones from poultry like chicken and duck get eighteen hours of cooking. Angie uses the bone broths in soups, stews, and sauces; as braising liquid; and as the liquid to cook rice or grains.

Bone broths, smoothies, meat protein, and fermented vegetables are the pillars of her diet for mothers during the crucial first few months of the postpartum period. That fermented squash, for example, is made from thin slices of peeled butternut winter squash with additions of ginger, garlic, onion, leek, sea salt, pink peppercorns, hot chiles, and anything else that takes her fancy except for the cabbage family members. "They go under the brine—it's important they stay covered with brine—for about six weeks, at least," Angie said. "I like them sour. And the ingredients need that time for the flavors to merge." She uses a strong brine of one tablespoon of sea salt to each cup of water. She checks the ferment and skims any yeast growth from the surface of the brine, and stirs up the fermenting squash once a week.

"I like sauerkrauts made from fennel—but everything can be fermented." She keeps her menus ethnically interesting and timely—she has a Greek menu, a Moroccan menu, a St. Paddy's Day menu, and so on.

"One important step for me is to sterilize the fermenting vessels and storage jars. I rinse the fermentation pots with boiling water before I add the ingredients," she said.

One of her interests is the Paleo diet—trying to eat the kinds of foods our hunter-gatherer ancestors ate before agriculture. That means pretty much eliminating dairy and breads. "I'm drawn to the diets of Native American people," she said. Most pre-Columbian tribes were very much agriculturists: witness the development of corn, chiles, tomatoes, and potatoes; the making of milpa in Central America; and the extensive fields of corn, beans, and squash (the three sisters) enjoyed by indigenous people throughout North America.

A typical MamaKai menu is twenty-three items, divided into "freezer friendly," "eat fresh quickly," and "longer shelf life" dishes. Each dish is given a full description. One of the ferments she offers is fennel herb kraut. Here's what the menu card says about it: "Lacto-fermented foods add lots of healthy and very beneficial bacteria to our guts, aiding in digestion, maintaining a healthy digestive tract, as well as boosting our immune systems as being the first line of defense against outside invaders. We wholeheartedly believe in eating a wide variety of lacto-fermented foods and beverages, and will be sure to have a rotating variety of seasonally fermented menu items to help you and your family stay extremely vibrant and healthy. Fennel herb kraut is made with onion, fennel, celery, celery root, garlic, chiles, rosemary, parsley, oregano, dill, seaweeds, sea salt, and filtered water."

Angie has also created a sixty-page companion guide to help educate women about the nutritive value of fermented and other foods. In stocking clients' refrigerators with live, nutrient-dense foods, she has devised a business that relieves new moms and moms-to-be from many of the anxieties that can overwhelm new families. Angie is just thrilled, she told me, to have found this path in life—one that truly makes a difference in people's lives, and one that relies on fermented foods as a central part of why and how MamaKai thrives.

A Visit to the Cultured Pickle Shop

Cultured is the name of the company that makes a wide range of fermented foods in Berkeley, and the Cultured Pickle Shop is its retail outlet. It's a fascinating place to visit because it gives you tons of ideas for what to culture with lactobacilli. In case you're not up on the process, there's a display on the north wall that reads "The Bacterial Progression of Lacto-Fermentation." Below the caption are four photos of bacteria in the order in which they colonize vegetables undergoing a lacto-fermentation. First in are the *Leuconostoc mesenteroides*, which get things ready for the growth of *Lactobacillus brevis*. These in turn prepare the ferment for the appearance of *Pediococcus cerevisiae*. Slowly the *Lactobacillus plantarum* begin to take over and eventually become the dominant fermenters.

Other progressions can and do happen, but this is a common, typical one. All are beneficial in the preservation of food by the conversion of sugars into lactic acid.

As you enter the Cultured Pickle Shop, there is a small space for customers, about the size of a six-by-nine jail cell, with a large cooler at the east end and a large chalkboard with the specials being offered. A sign that reads "Employees Only Past This Point" stops you almost as soon as you enter. And that's understandable. In the large inner room, employees man (and woman) the tables, preparing the ferments, and it wouldn't do to have customers wandering about.

The cooler is stocked with familiar fare—many kinds of sauerkraut, fizzy kombucha, kimchi, and so on. And here's what was being offered on the specials board on a recent weekday:

Kimchi
—Watermelon rind
—Celery and cauliflower
—Pumpkin and bok choy
—Mustard greens and leeks

Tsukemono (Japanese for pickle)

—Kasu-zuki (burdock and cucumber)

—Miso-zuke (burdock and turnip)

—Nuka pickles

Kombucha

Year-round availability

—Celery

—Beet

—Fennel

Seasonal availability

—Tangerine-turnip

—Grapefruit-daikon

—Asian pear–peppercorn

—Pumpkin-peppercorn

—Turmeric-ginger

Seasonal Specialty Ferments

—Beet and fennel

—Red daikon with chrysanthemum greens, jalapeño, and shiso

—Beets and celery with dill and juniper

—Spicy oregano purple carrots

—Radish and fennel with bee pollen

—Celery, sunchoke, and radish with jalapeño and coriander

—Pumpkin with green onion and espelette peppers

—Classic or spicy dill pickles

—Indian pickles and limes

—Moroccan preserved lemons

—Chili paste

—Tokyo cross turnip with turmeric brine

Berkeley is a wonderful community, always a step ahead, and here Cultured is offering the community what heretofore had been available only to home fermenters. It's very much like the availability of organic food in the 1940s through the 1960s—that is, if you wanted it, you pretty much had to grow it yourself. Home fermenting is so easy and fun, though—and so magical as you watch your vegetables, milks, juices, and flours turn into nutritional powerhouses—that I'm hoping it remains a home-based movement, and not just a slogan on a cooler at Whole Foods.

As you can see from the specials board, there is no set recipe for fermenting vegetables, but rather a technique, with the ingredients up to what sounds good to you. The Cultured Pickle Shop does commercially exactly what you do at home to make your own cultured vegetables and drinks, and that is to submerge the prepared veggies in brine to encourage the development of lactobacilli and other beneficial microorganisms. What the Cultured Pickle Shop offers residents of Berkeley is not anything different from what you make at home, but rather it offers convenience for busy folks who may not have the time to ferment their own edibles.

PART 2

A Fine Meal of Bread, Cheese, and Wine

In this section, we'll delve deeply into three fermented foods familiar to everyone that together make up a triumvirate of deliciousness. We'll visit a breadmaker at work, a cheesemaker at his stainless-steel tubs of sheep's milk, and a home winemaker who's decided that homemade bread, cheese, and wine can be nature's finest foods.

CHAPTER 8

Bread

Is there an aroma more mouthwatering than that of yeasted bread baking in the oven? The scent sums up and integrates the ingredients of great bread: bread yeast, wholesome flour, and spring water. Of these ingredients, the one that makes great bread, not just good bread, is the flour. So in this section, we'll be examining flour in great detail. The more you know about your flour, the better your bread will be. Of course bread is not a source of live bacteria and yeast—they die off during the bake—but it does contain the metabolic products of the yeasts and/or bacteria that leavened it. Slather it with cultured (fermented) butter, and you are back in probiotic business.

Probably no change in our culture sums up the emergence of the organic ideal more than the change in our most familiar fermented product, bread. Back in the mid-nineteenth century and earlier, one took one's grain to the mill and the miller ground it into whole grain flour. As long ago as ancient Rome, the upper classes preferred whiter bread, but it wasn't until the second half of the nineteenth century and the advent of steel roller mills that supplanted millstones that the wheat berries were reduced to simple starch. The baking industry had finally learned to take the bran and germ from the wheat, giving us the tasteless, gummy, white slices used to hold sandwich ingredients together. It was a utilitarian product, not a palate pleaser. It's hard to believe we ever got along without the wonderful breads we've rediscovered since the 1960s, made the way breads in Europe were made: pain au levain, made with natural starter; good baguettes that must be eaten fresh from the oven to be at their best; ciabatta, the Italian word for "slipper," which is a slack-dough bread with an airy, soft interior and a light, thin crust; quintessentially German dreikornbrot, packed with grains and seeds and bursting with natural goodness; and many other types of fine bread.

My thoughts ran along those lines when I recently brought home a loaf of sourdough from the Acme Bread Company of Berkeley, California, and tore—literally—into its dark brown crust. I ripped off a chunk of crust with my side teeth and crunched in. It had yielding parts, tough and chewy parts, and crisp and crunchy parts with a browned and roasted nutty wheat flavor. Inside the bread was moist, stretchy, and chewy. Its aroma combined a dominant sourness with a satisfying, warm, and yeasty graininess. The smell of this bread brought back memories of afternoons when I was about ten, climbing into the beams of the neighbor's dairy barn, jumping off into the dry hay, and smelling dusty alfalfa and timothy, oats and chaff, fermenting corn silage, milk and cows, all mingled together. But this Acme loaf is not the bread of my childhood—not by a long shot.

"There's a big change going on with bread," says Joseph Rodriguez of Uprising Bread Bakery in Brooklyn. "The top bakers in Paris are going organic. True artisans all over the world are sticking to pure, organic ingredients." Uprising is an organic bakery started by Rodriguez and Nicole Lane, his wife. The bakery, with two retail outlets plus sales to markets and restaurants in New York City, makes twelve or thirteen kinds of naturally leavened breads daily, including Italian country bread, special breads like potato and rosemary and caramelized onion, ciabatta, and French-style baguettes. "We don't use commercial yeast," Rodriguez says. "We've created a natural starter culture and refresh it often—it gives a slightly sour flavor to the bread, but not as sour as San Francisco sourdough."

Natural Starters and Commercial Yeast

Natural starters are made by simply letting a mixture of flour and water ferment through the agency of whatever yeast and bacteria are floating in the air. Because each place on the face of the earth has a unique mixture of microorganisms, each place can produce a starter culture that's unique to that place. That's one reason a natural, artisanal bread can show *terroir*, or the taste of the place it's from. In San Francisco, for instance, the sourdough agents are a pair of microorganisms—the bacteria *Lactobacillus sanfranciscensis*, which is native to the region, and the yeast *Candida milleri*. They propagate together in the dough, working symbiotically to produce that unique, very sour flavor of San Francisco sourdough. *L. sanfranciscensis* produces the sourness by utilizing the maltose sugars in the dough. *C. milleri* uses glucose sugars to produce alcohol and the carbon dioxide that leavens the bread, making it rise as the glutinous dough fills with expanding gas bubbles. Thus, there's no competition for a nutrient source between the two. Commercial yeast (*Saccharomyces cerevisiae*), however, also uses maltose, which makes it compete with *L. sanfranciscensis* and interrupt the production of the souring compounds. That's why a natural leaven produces a more idiosyncratic sourdough bread than commercial yeast.

For the recipe for making your own natural starter, see pages 223–25.

Rodriguez says that he and Lane are proud to be a part of the organic bread movement. "Here in the United States and around the world, it's growing by leaps and bounds," he says. All Uprising's flour and other ingredients such as oats, onions, and apricots are organic, "except for nuts," Rodriguez says. "Organic nuts cost twice as much as conventional nuts, and we just can't afford them. But all the coffee we serve in our stores is organic, and all our spreads and jams. We push organics."

Why?

The answer is as much social as it is agricultural. "I grew up in Kansas and met Nicole when she was going to the University of Wisconsin at Madison. We got involved in the organic community, which is really strong in Madison. We used to work in the community gardens." After Madison, Rodriguez and Lane studied at the San Francisco Baking Institute, a school that has graduated many bakers with a strong commitment to organics. One of their instructors was Lionel Vatinet, a French master baker who now operates La Farm Bakery in Raleigh, North Carolina, and who also uses organic flours. Lionel has been instrumental in educating many fine artisanal bakers in the United States, including Kathleen Weber of Della Fattoria Bakery in Petaluma, California, whom you will meet later.

Vatinet says he likes to use organic flour because of the nutritional punch it gives. "Ten years ago, most bakeries used commercial flour that was bleached, bromated, and full of chemicals used to fortify the flour nutritionally. But I'm from France, as you can tell [spoken in his liquid French accent], and we work with unprocessed flour over there. With organic flour, especially whole wheat, there's enough nutrition that a human being can almost live on it alone," he says.

"Identity Preserved" Bread Flour

At Uprising in Brooklyn, Rodriguez and Lane—along with many other organic artisanal bakers around the country—use flour from Cook Natural Products of Oakland, California. Cook Natural Products is one of those enabling companies whose work foments and promotes the blossoming of organic bakeries. Without Cook, or a company like it, artisanal bakers would have a much harder time securing reliable supplies. Its director of Technical Services is Ian Duffy, a former rock climber, cyclist, tofu maker, baker, and instructor at the San Francisco Baking Institute. He was born in Munich, Germany, and raised on delicious Bavarian rye before moving to the United States at about age ten, where the lack of good bread made the move more difficult than it had to be.

"More than half of the flour we sell is organic," Duffy says, "because selling organic flour is the right thing to do." The company prides itself on its line of organic flours and other grains, and sells conventional flours and grains as well. "I also prefer the sensibilities of our organic customers. And we can't compete with ConAgra [a major supplier of conventional commodity flour to the big baking companies of America], so organic is a niche market for us."

Duffy explained why Cook is unique in the way it secures organic wheat, mills it, and sells it to artisan bakers. "The big conventional wheat growers sell to brokers and the brokers sell it to the big grain suppliers; it operates like the stock market," he says. "Or they sell to grain elevators at a guaranteed price and the elevators sell to ConAgra and other big wheat wholesalers.

"Since elevator wheat comes from many different sources, it varies in moisture content, weight, and protein content—all important numbers for bakers to know. So millers blend different lots together until they get the specifications they want. They may be hitting consistent specs, but because the blends are from always-changing components, any unique characteristics of the wheat are lost. It's the wheat equivalent of jug wine," he says.

"Our idea here at Cook is more like varietal wine. We select certain kinds of wheat we know have definite characteristics like high protein content, and we ask farmers to grow these selected wheats. Then we contract with mills around the country to make flour from them," Duffy says. Some of these mills are roller mills (using large steel rollers that crush and grind the wheat to fine flour) and some are stone mills (where wheat is ground by huge revolving millstones held a hair apart). Some bakers prefer stone-ground flour because they believe it's ground at cooler temperatures and more of the enzymes and volatile flavor compounds in the wheat are retained, but there's some question as to whether this is accurate.

"Then we ship our flour from these contract mills to nearby customers," Duffy says. It's a marketing tactic, because "a lot of the cost of flour to the baker is not in the growing or the milling, but in the moving and storing, especially with organic flour." With organic flour, storage bins and trucks have to be thoroughly cleaned of conventional flour before organic is put in, and that costs more. Trucking is expensive. So the closer the mill to the farmer and the baker, the fresher the flour and the lower the cost.

"For every lot of wheat we mill, we do lab tests to show how the flour will perform. When a baker buys flour from us, he or she knows the lot number and can look on our website to see the lab reports. The more artisan bakers know about their flour, the better for them because they can use that knowledge to bake the best possible bread," Duffy says. Over a dozen different tests are performed on each lot of wheat and the flour made from it, including tests for protein and moisture content of both wheat berries and flour, hardness of the kernels, wheat color, ash content of the flour, and falling number. The falling number test determines the enzyme activity of a flour sample. The test entails heating measured amounts of water and flour in a special tube. The tube is placed in a boiling water bath and stirred with a plunger until the sample is gelatinized. Then the plunger is placed on the surface of the sample, and the time that it takes the plunger to sink to the bottom of the tube is recorded. Depending on the enzyme activity, the degradation of the starch paste will vary. The higher the enzyme activity, the more the flour is degraded and the faster the plunger reaches the bottom. Generally the baker will find that fermentation progresses more rapidly as falling numbers become lower.

These test results are posted as to lot number, so bakers really know what they're dealing with. Bakers also find out which wheat variety is used in their flour, where the wheat is grown, the name of the farmer who grew it, and how long the lot will be in stock before it's replaced by a new lot.

"In the past, the problem with organic flour had been its variability," Duffy says. "It's usually produced by small companies with small mills and small storage facilities. So a baker buys flour grown by farmer X this week, and next week it's another lot, grown by farmer Y. But lots vary tremendously, and this causes problems for bakers who want consistency in their high-quality loaves. Here at Cook, half of our entire wheat supply for six months is purchased and stored already, so we get more consistency, and each lot is numbered under our identity preservation program so the baker can check its characteristics and track it from farm to bakery."

Besides the baker, the farmer benefits from the identity preservation program.

"Small farmers, especially organic farmers, can't make money growing wheat and selling it through brokers or to elevators. Wheat is a commodity with prices set so low the farmer gets beat up on price." Huge factory farms can do it because they have economies of scale and are sometimes vertically integrated with big grain companies like Archer Daniels Midland and ConAgra. "But selling organic wheat is another story," Duffy says. "Organic farmers are paid more. A hundredweight of conventional, commoditized wheat sells for about $5 at the farm. The organic farmer gets a 150 percent premium—up to $12.50 a hundredweight." Cook thus has set itself up as a supplier to small artisan bakers looking for flour of a quality commensurate with their talent, bakers who want to produce bread with *terroir*.

An Organic Flour Supplier

One of the chief suppliers of organic flour to the artisan baking trade is Giusto's Vita-Grain in San Francisco. "In 1940," says Keith Giusto, the current owner, "my grandfather bought a whole grain bakery and health food store on Polk Street, and made a nine-grain bread. He milled his own wheat for it."

(In an aside, he says his grandfather was captain of the country's first professional soccer team. Maybe it was that good nine-grain bread that gave him the energy.)

"People would ask him, 'How come your bread tastes so good?' and they would offer to buy his flour. So he started selling flour and it became a big part of his business. In 1960, my dad and uncle found organic grain farmers and started selling organic flour—so we've been doing it for a long time."

Keith is the third generation of Giustos in the business. He's almost fifty years old and still hunting for new sources of organic wheat. "Tonight I'm going to Washington [state] to check on a new organic farmer," he says. "With organics, you have to be a good farmer, and you have to be planting the right varieties. Our most popular flour is made from 100 percent winter wheat with a protein content of 10 to 11 percent. That's just about right for artisan bread."

Besides the organic flour business, Giusto runs a baking school in Petaluma, about an hour north of San Francisco. "People come from all over, some as far as Arizona, to attend these classes. They're small, four- to six-person custom classes, and most of the students are professional bakers who want to learn more. We'll educate you as deeply as you want to be educated about the whole process, from farm to mill to bakery," Giusto says.

Giusto is in a tight relationship with his organic farmers. "We're loyal to our farmers and they're loyal to us," he says. "Try that with ConAgra and they'll tear you up. For instance, five years ago, we started helping farmers find outlets for their rotation crops of corn and soybeans." Organic farmers, as opposed to most conventional ones who use chemicals, rotate crops. They'll follow wheat with soybeans, a legume that puts nitrogen and organic matter into the soil. The following year, there's enough nutrition to bring in a crop of corn. Cornstalks are plowed under, and their decaying organic matter is sufficient to support a crop of wheat in the fourth year. The depleted soil is then revitalized with a crop of soybeans the following year, and the cycle begins again. This is sustainable farming. Most conventional farms, on the other hand, flood the land with chemicals to support the same crop year after year. These chemicals can destroy the microorganisms that dismantle crop residues into nutrients for subsequent crops to use. As one farmer who year after year plants "continuous corn" told me, "I plow the cornstalks down in the fall and plow them back up in the spring." Because of the lack of life in the soil, the cornstalks don't decay to feed the soil. Small organic farmers cannot guarantee buyers full production of wheat each year because of their need to rotate crops, and to make up the slack in wheat income, they need to sell their corn and beans. Giusto's ability to find markets for their rotation crops helps them immensely.

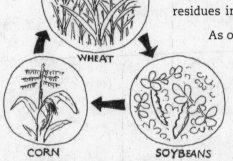

Typical crop rotation

The Taste of Maine in a Loaf of Bread

That's a concept dear to the heart of organic baker Jim Amaral and his wife, Dolores Carbonneau, who in 1993 started Borealis Breads in the basement of the Pine Cone Tavern in Waldoboro, Maine. Interestingly, Amaral was a winemaker at Sakonnet Vineyards in Rhode Island, where he grew Pinot Noir, Cabernet Sauvignon, Cabernet Franc, and a white French-American hybrid called Vidal before starting Borealis Breads. The move from wine to bread isn't that great a stretch, considering that both are fermentation products that are at their best when they're locally produced and organic.

Amaral started with just twelve breads and twelve accounts. Today the business employs fifty-five people and annually buys a million pounds of flour that go into 1,440,000 loaves of bread baked at four retail and two wholesale bakery sites around Maine. He's looking forward to the day "when we will have a range of breads made exclusively from Maine wheat varieties bred by Maine grain breeders. The wheat will be grown in Maine. The bread will be unique to Maine, using Maine techniques, Maine varieties, natural leavens from Maine bacteria and yeast, and all local resources," he says. Of course, he's talking about *terroir*. "I want people to eat my bread and say, 'That's Maine wheat,'" he says, just as people now associate certain tastes of wine and other foodstuffs with particular regions of Europe or California.

The idea of locally grown and baked bread is one of the key reasons why Borealis Breads is so successful—that, and being organic. "People aren't just buying the bread, they're buying the company," Amaral says. "If I'm using local ingredients, people will buy that. Organic is important, too, but honestly, locally grown is most important and organic is second to that." In addition, Russ Libby, the executive director of the Maine Organic Farmers and Gardeners Association (MOFGA), encouraged

A real farmer you can call up on the phone or reach via e-mail

Amaral to "put a face on the product." Amaral took heed and, with a $10,000 grant from the Maine Department of Agriculture, produced a series of "farmer cards," similar to baseball cards but with photos of individual Maine farmers and recipes on the back. "We hand them out with every retail transaction," he says. The cards have become wildly popular. I can see why. Brand loyalty is something the big agribusiness companies know all about. But when the brand is also a person, it's an even stronger marketing concept. Look at Orville Redenbacher, Oscar Meyer, Betty Crocker, Colonel Sanders. When the person is a real farmer you can call up on the phone or reach via e-mail, brand loyalty becomes almost a family affair.

It hasn't always been a million loaves and farmer cards, however. "When I started out," Amaral says, "I began with four different organic loaves. People weren't buying them. It was difficult. But that's turned around. People now are paying attention to whether bread is organic or not. I give lots of credit to MOFGA and their leadership, but what has really turned the market around is the media attention on organics when the USDA issued its organic rules, and the fact that people know that organic on the label means there will be no genetically modified ingredients. Farmers—they're terrified of GMOs. It could kill their overseas business."

The process of achieving the goal of producing locally grown organic bread is well under way. Aroostook County in northern Maine, known these days primarily as a source of fine potatoes, "was the breadbasket of New England during the Civil War," Amaral says. "We're working with farmers there to grow hard red winter wheat with a protein content of 12.5 to 15 percent." That high protein percentage is characteristic of wheat grown in northerly areas—a reason the northern part of the Great Plains of the United States and Canada is now the breadbasket of the world. When protein in flour is hydrated with water and kneaded, it forms gluten. The higher the percentage of protein in flour (or the "stronger" the flour, as it's called), the more gluten can be formed. Gluten is the substance that makes bread elastic, stretchy, and able to swell with the gases released during fermentation, resulting in light, chewy breads with many air holes. Very high protein wheat—over 14 percent or so—can result in strong white flour with too much gluten, producing breads that are rubbery, dense, and tight. Wheat grown in more southerly climates has from 8 to 10 percent protein or less and is good for pastry, waffles, pancakes, cakes, and biscuits, but doesn't have enough gluten to react to the gaseousness of fermentation by producing light, chewy bread. That's why the baking powder griddle cake is associated with the South and the loaf of yeast-risen wheat or rye bread is associated with the North. All-purpose flour is so labeled because it's designed to be good for all these kinds of uses and more. According to many sources, something like 11.5 percent protein is ideal in a white, unbleached organic flour destined to make ideal white bread. Whole wheat, on the other hand, can be stronger, with about 14 or 15 percent protein. Whole wheat contains everything in the wheat berry—the bran, which is the fibrous outer coating of the berry; the germ, which is the vitamin E–containing, nutritionally power-packed site of the embryo that would become a new wheat plant if sown; and endosperm (the white starchy part of the seed). When flour is milled from whole wheat, the tough bran is broken into tiny fragments that can slice through the long strands of gluten, deflating the air holes that make bread light, and creating a dense loaf. Therefore the extra protein in whole wheat is beneficial, for it will create more gluten that will overcome the effect of the bran.

Grains other than wheat don't produce gluten, or very little. That's why rye, barley, oat, and rice flours are always mixed with wheat flour—unless a dense flatbread is wanted.

Bread Boffo on Broadway

Amy Scherber began with a small bakery in the Hell's Kitchen neighborhood on the West Side of Manhattan. Now just about everyone in New York City knows Amy's Bread. There are three retail locations in Manhattan and her bread is sold in many stores around the city. From the beginning, she was determinedly organic, but it wasn't easy.

"Organic hasn't always been an easy sale," she says. "Not all the customers want to pay extra for organic. Or they don't understand what it is, so they think it sounds like health food, which they avoid. In New York, we are way behind the

West Coast on the popularity and availability of organic products, but we have slowly developed a following for our products because they taste great and are organic, too."

Given the difficulties, Amy could have made conventional bread, but "I like to use organic flour when I bake because I feel it has a positive, long-term effect on the soil used to grow the wheat, and on the sustainability of the land and the groundwater." But taste enters into it also. "We make some delicious organic breads that are not sour or tangy," she says. "I prefer a mild, earthy, wheaty, non-sour bread with my cheese, so that's why I choose organic. I get those results with my organic flour."

She buys her flour from Cook Natural Products, Giusto's Vita-Grain, and Community Mill & Bean in upstate New York. But Amy sounds a bit dismayed when she says, "None of my customers have ever asked where the wheat or flour come from. I hate to admit it, but it's true."

A Visit to a Small Bakery

Country boy Lou Preston has made a name for himself as a winemaker. Preston wines are grown and vinified on Lou's 125-acre ranch at the northern end of Dry Creek Valley, which runs northwest from the town of Healdsburg in Sonoma County, California. He started making wine at home in the 1960s and bought the ranch in 1973. He's had a bonded winery, Preston Vineyards, since 1975.

At first he didn't use insecticides, but soon got into a spray program that snowballed. "I found I was spraying more and more chemicals in the vineyards," he says. "In the mid-1980s, I went cold turkey and stopped spraying," and now he has forty-five acres of grapes certified organic by California Certified Organic Farmers (CCOF). "The reason I went organic is personal, it's selfish. I didn't want to be surrounded by a toxic chemical environment," he says.

Recently I heard rumors that he was getting out of the wine business and devoting himself to baking artisanal loaves of organic bread. That's not exactly right, I discovered. He's reorganizing his wine business by downsizing it from 30,000 cases a year of thirteen varieties to 8,000 cases of seven varieties—and baking more natural-starter organic bread. "People who've gone the extra mile to be organic obviously care about their product. So customers know that the product is good and that the farming techniques used to grow the raw materials protect the environment," he says. "The reorganization has been difficult," he admits, but to watch the ease, skill, and enthusiasm with which he bakes shows that for him, it's been worth it. Part of the reorganization has been to diversify his ranch. He's planted a thousand olive trees for oil and cured olives, and is looking for other crops that host beneficial insects, are aesthetically pleasing, appropriate for Dry Creek Valley's hot day–cool night climate, and add biodiversity to the ranch. "I'm working now to see if I can reproduce the Tabasco process," he says. "If I get a lactic fermentation going in the peppers, the pH of the sauce goes from 5.0 to 4.0—and I read that pHs under 4.7 prevent the botulism toxin from forming."

LET THE BAKE BEGIN

I showed up at Lou's ranch at 6:30 on a recent Saturday morning to watch him bake, and found that I got there a few minutes before he arrived. I wandered a bit in the back where there was an old concrete kiva-style bake oven and an extensive organic garden with corn, squash, sunflowers, figs, cardoons, and much more, all patrolled by chickens clucking happily in the warm, early morning sunlight. Two kittens about ten weeks old came up to me, hoping I had a morning dish of milk for them.

Then Lou arrived and we went into his baking room, an extension off the winery's tasting room, dominated by a beautiful masonry and brick oven built by the late Alan Scott.

Scott was an interesting character, and the one person who is perhaps most responsible for fine, small-scale bread baking in Northern California and around the country, for it was he who built the masonry and brick hearth ovens that he believed would become the centers of new communities of like-minded people.

These communities are organically oriented, health conscious, and rooted in an environmental ethic that values the handmade, local, and artisanal over the cheap, mass-produced, and globalized.

"This book," he wrote in *The Bread Builders,* "contains heaps of my enthusiasm for . . . those true baker-artisans . . . who are now successful family and community nurturers. Without nurture I do not think there can be nutrition, since nutrients, numbers, and other heady stuff can lack heart, whereas nurture, being from the heart, is the more powerful mover and shaker." He believed that a fine bakery serves "our families, friends, and communities." The book not only describes the ovens Alan built, but provides plans in case you're ready to do backyard hearth oven baking yourself.

The oven Scott built for Lou Preston is large, with a four-by-eight-foot baking floor. "It takes about a square foot of space per loaf, so I can bake almost three dozen loaves at a time," Lou says.

He peeked in the oven where a wood fire had just about burned itself out. The Saturday bake starts on Thursday, when the oven has cooled to about 300°F from the previous bake on Tuesday. (He bakes twice a week.) Lou loads the oven then with about two and a half wheelbarrows full of oak, but doesn't set it afire. It spends a day roasting in the oven so it's nice and dry for Friday, when it's set ablaze. The wood burns all day Friday, the oven temperature rising to about 700°F. By Saturday, the fire has died down and the oven is about 475 to 500°F, just right for baking.

The organic flour—white, whole wheat, and rye—is from Giusto's Vita-Grain. On the previous day, Lou had made dough for our bake out of thirty-seven pounds of white unbleached flour and three pounds of a mixture of whole wheat and rye. To this he added twelve pounds of his wet natural starter, plus enough water, about twenty-eight pounds, to make a fairly slack dough that's 74 percent water. Almost all bakers measure their ingredients by weight. It's more accurate than by volume, and accuracy leads to consistency. "I make a sourdough country white bread from this," Lou says. He'd shaped the dough into loaves and placed them in the folds of a couche—a baker's linen—to proof, as the rising process is known in baker's lingo.

Since he filled the folds of the couche on Friday, they would overproof if allowed to sit out for a whole day, so he placed them in his walk-in cooler to retard the rising. An overproofed loaf has spent its yeast and carbon dioxide gas and won't rise properly. There's a bonus to keeping the dough in the cooler, too: "It develops flavor."

HANDLING THE NATURAL STARTER

A natural starter is an active mix of dough and living microbes, mostly yeast, instead of the dry commercial yeast that most bakers use to leaven their bread. Because the ferment starts naturally in a mix of flour and water, it reflects the microbial population in the air of the bakery and gives the bread more of a taste of the place it's from. Lou's had his natural starter for ten years. "I learned about starters at the Culinary Institute of America [in the Napa Valley]. For an overnight starter, we used to put grapes into the flour and water so the yeast on the skins would colonize the starter and begin to ferment it. I later learned that natural starters take on the characteristics of where you make them and what you feed them. It takes months to get a starter of natural yeast and bacteria up and working strongly. Over that time, some strains die out, others grow stronger, until you get a stable mix of diverse fermentation organisms." Interestingly, the Acme Bread Company in Berkeley uses a natural leaven that it has had for twelve years and that was begun with the yeast and bacteria on grapes—possibly from the same class at the Culinary Institute that Lou attended.

I watched, fascinated, as Lou cleaned out the floor of his masonry oven. First he scraped out the bulk of the wood ashes with a flat board at the end of a long handle. Then he scrubbed out the floor with a wet push broom, and finally mopped the floor with a wet rag on the end of a long pole. It all seemed so familiar, so ancient as well as modern, and I realized that if I could be transported back in time a thousand years or more, I'd be able to witness pretty much the same procedures being carried out.

Lou told me how he keeps the starter going. When he makes dough, each batch takes twelve pounds of starter, or 15 percent of the total dough weight of about eighty pounds. He needs to save four pounds as the seed for the next bake.

That means he needs to begin with sixteen pounds of starter. Starters need constant refreshing, and so to the four pounds of seed starter he adds enough flour and water to double it to eight pounds. The day he makes dough, he mixes this with four pounds of flour and four pounds of water, giving him sixteen pounds of starter. He takes twelve pounds for the bake and reserves four pounds to be the seed for the next bake. And

Lou Preston's starter procedure

so it goes, on and on. There are people in San Francisco who claim their sourdough starters have been going since the Gold Rush. But so what? As long as there's *Lactobacillus sanfranciscensis* and *Candida milleri* in the air, anyone in San Francisco can make a sourdough starter that tastes the same as a starter that's been going for 150 years. And starters can be made naturally anywhere in the country, not just the Bay Area—although San Francisco is known for the extra acidic and sharp quality of its sourdough bread.

GETTING THE DOUGH OVEN-READY

Lou wet-mopped the oven again and sprayed it with a hose with a long brass rod and nozzle that reaches all the way to the back. "This mitigates the temperature back there so it's not so hot that it kills off the yeast in the dough before it's had a chance to inflate the bread," he says.

Now he takes a metal tray with five loaves of unbaked dough nestled in the folds of the couche and pulls the ends of the couche so it flattens out and there's space between the loaves. He has two one-by-six-inch boards, one in each hand. With his left hand, he places the board vertically next to one of the loaves and expertly flips it to the right onto the board in his right hand, which is held horizontally. This he holds level and slides the dough with a quick, jerky motion onto a peel—the long-handled, flat-bladed wooden utensil used to slide loaves into and out of the oven. When he has all five loaves lined up on the blade of the peel, he reaches for his lame (pronounced *lamm*), which is a tool for slashing the top of the bread. It's about five inches long and holds a curved razor blade that slips under the surface of the loaf as it slashes. Sometimes the dough will puff out through these slashes before it's even placed in the oven. Usually, however, when the dough heats in the oven, the yeast furiously makes gas, which expands the bread up through the slashes. This allows for good expansion and has the added bonus of giving the bread a crustier surface area. One can use a sharp serrated knife, but I find the lame is the perfect tool for the job. Lou dips it in water between loaves to keep the razor edge clean. Elongated Vienna-type loaves are slashed with two slightly oblique slashes. Round loaves are slashed with three slashes that cross in the center, giving a six-pointed effect. Other round loaves are given a semicircular slash "so that the top pops up like a little hat," Lou says. Once the loaves are slashed, he walks them to the oven, opens the door, slides them into the back, sets the peel aside, sprays the inside with the nozzle again (the steam helps the loaves achieve a crunchy, perfect crust), and repeats the process until all the loaves are in the oven. The loaves get steam only during the first ten minutes or so of the bake.

PAR-BAKED AND FULLY BAKED LOAVES

The loaves go into the oven at 7:30 a.m., and Lou says they'll be finished in forty-five minutes. Because his winery visitors often buy loaves, and they are only oven-fresh on Tuesdays and Saturdays, he par-bakes some of his loaves, a method he learned from Nancy Silverton at the La Brea Bakery in Los Angeles. "You take the par-baked loaves out at twenty-five minutes and cool, then freeze them. When you get them home, pop them into a 450 oven for twenty or twenty-five minutes, until they look right and sound hollow when thumped on the bottom. You can hardly tell the difference from once-baked loaves."

A number of Lou's loaves come out at twenty-five minutes for cooling and freezing. He takes the peel and slides the loaves remaining in the oven into positions he knows from long experience will be the right places for them to finish a perfect bake. After a while, he uses the peel to retrieve the fresh, crackling loaves from the oven. They are a dark brown with all sorts of color variations in the crust and slashes that have expanded. "I like the darker-colored crust rather than a light one," he says. "If I had more dough ready, this oven would be perfect right now. The second bake is usually the best."

He selects a loaf and we tear it open, still steaming hot inside. It's lightly sour ("It would be more sour if I took more time to let the starter develop after refreshing it," he says) and, as I find out after I get it home, acquires an even sharper sour flavor when it cools. The flavor is incomparable, the crust a jingly, crunchy, nutty joy. The bread inside the loaf is hot and stretchy, still smelling of yeast and wheat. Lou hangs up his peel, and as the loaves sit on trays to cool, I can hear little ticking sounds as the crusts crack as they cool.

My mind is on the bread, which I know will make a perfect lunch when I spread it with runny, tangy Teleme cheese and eat it with a glass of Lou's spicy, tart Barbera wine.

More About Natural Starters

Lou Preston's fermentation added layers upon layers of flavors to his bread. His natural starter is now, after all these years, a carefully balanced ecosystem of yeast and bacteria, but to understand the sourness of his bread, I did some research into the bacteria that accompany yeast in starters. All starters, even a packet of commercial dried yeast, have bacteria associated with the yeast, although in a commercial packet of dried bread yeast, it's probably not more than a few percent. Natural starters have much more bacteria, and thus tend to make sourdough-type breads. The bacteria are those lactobacilli that produce lactic acid, just as they do in cheese and yogurt, and *Acetobacter*, which converts the alcohol produced by yeast during the fermentation into acetic acid, the sharply sour kind of acidity one detects in vinegar.

Acidity strengthens the gluten in bread, and you'll notice that sourdough breads and those made with natural starters have a tougher and more elastic texture than breads made with commercial yeast. That's because the natural starter produces more acid, and the longer it sits without being refreshed, the more acidic it becomes. Too much acidity in the starter could be a problem for a baker like Lou, who only bakes twice a week, and so he keeps his four pounds of seed starter in the fridge to slow it down. In large sourdough bakeries producing hundreds and thousands of loaves a day—each needing a big dollop of starter—the natural leaven is refreshed several times a day.

Once the starter is introduced into fresh flour and water to make dough, it starts to produce acids. If the fermentation is cool and drawn out, *Acetobacter* is favored and the bread acquires a sharper, more vinegary flavor. If the fermentation is warm and relatively quick, lactobacilli are favored and the bread acquires a softer type of acidity with a milky flavor. Lionel Vatinet of La Farm Bakery in North Carolina lived in San Francisco for years when he was an instructor at the San Francisco Baking Institute, and he says, "The typical San Francisco sourdough is too sour for me. It has too much acetic acid and not enough lactic. I prefer the flavor lactic acid gives to bread." He and many other professional bakers (except for those dedicated sourdough aficionados in San Francisco) strive to achieve a fermentation at warmer temperatures so that lactic acid predominates over acetic. It's not that acetic acid is bad—it's one of the substances that make for complex flavors in bread—but rather that it should be subordinate to lactic acid to produce the most pleasing type and level of acidity, these bakers feel.

Another factor in the mix of yeast and bacteria found in a natural starter is the flour. Conventional flours are treated with chemicals to bleach them and preservatives to prevent any microorganisms from colonizing them, to say nothing of the fumigation they get in grain storage and further chemicalization of the bread made from them. Organic grain, on the other hand, comes equipped with a strong, natural mixture of yeast and bacilli that the wheat berries acquired on the organic farm and in subsequent handling on the way to the miller. The milling process, especially stone grinding, spreads these microscopic bits of life through the flour and adds more bacteria from the stone and the millhouse atmosphere. All those microorganisms contribute to the complex layers of flavor that organic breads reveal to the careful palate. In the organic scheme of things, microorganisms are not only good, they are essential for life and its quality. Yes, there are the occasional oddballs and rogues that cause problems, but that's as true of people as it is of bacteria. The answer is not to kill off all the oddballs (who'd want a world without them?), but to overwhelm them with the sheer number and diversity of the good guys.

One of the most influential voices giving advice for baking good, wholesome natural bread is that of Professor Raymond Calvel of the French milling school École Française de Meunerie. In 1990, he published *Le Goût du Pain,* which was published in English in 2001 as *The Taste of Bread.* While he believed that a lightly acidic natural starter bread could be quite good, he felt that its sourness limited its culinary usefulness and he preferred bread made with just a little bit of commercial yeast allowed to work in a very wet dough called a poolish for several days. This results in just a mild tang. He would also let freshly made dough of flour and water rest for a period of about an hour before adding it to the poolish on baking day. He called this rest period the autolyze. Along with many artisan bakers in the United States, I find much that's instructive in Professor Calvel's work, and I've included several of his ideas in my recipe for bread beginning on page 225. Another influential voice is that of the late Elizabeth David, author of *English Bread and Yeast Cookery.* Her charming book describes ancient, medieval, and modern methods of bread baking—all of which help to inform the baker about what's possible using the simple ingredients of flour, salt, yeast, and water.

Simple Flour Is a Complex Subject

That simple ingredient—flour—is not so simple once you start examining it.

Some varieties of wheat produce slightly yellow flour, and Professor Calvel claimed these have more carotene and more flavor than pure white flour from hard red winter wheat or hard spring wheat.

Spelt is a European grain that's a species of wheat (triticum) with slightly more protein. It makes delicious, mellow, nutty bread, and some people with wheat allergies can eat it.

Triticale is a cross between wheat (genus *Triticum*) and rye (genus *Secale*); it has less gluten than wheat but more protein and thus must be combined with regular wheat flour to create moist bread with a stretchy crumb structure.

Flavor is also dependent on location. For forty years, Arrowhead Mills has sold organic flour grown in Deaf Smith County, Texas, that's known for its intensified wheat flavors. This wheat is grown at an altitude of 3,500 to 4,000 feet, so the grain develops the same kind of protein levels—12 to 13.5 percent, according to Arrowhead's director of technical services, Dr. James Glueck—as hard wheat grown farther north. And, Dr. Glueck says, when supplies of Deaf Smith County wheat aren't sufficient to meet demand, they're blended with wheat from the wheat belt of Kansas and Nebraska. Arrowhead's sixteen types of organic flours are distributed nationwide through its parent company, the Hain Celestial Group.

Some of the lower protein flours grown in Europe and the southern part of the United States have incomparable flavor, but need additions of hard, high-protein wheat to strengthen them. In both Europe and America, a small amount of vitamin C (ascorbic acid) is sometimes added to low-protein flours to strengthen them, as acidity increases flour's ability to develop gluten from its proteins. Some yeast packagers also add a small amount of ascorbic acid to their yeast in order to boost its rising power.

An Organic Wheat Farmer

One of Giusto's Vita-Grain organic farmers is Richard Grover of Brigham City, Utah. He farms between 1,600 and 2,000 acres each year in nearby Snowville, Utah, just south of the Idaho border. I mentioned that I'd worked on *Organic Gardening & Farming* magazine in the 1970s, and was gratified to hear him say, "*Organic Gardening* was my bible back then."

He grows hard red winter wheat of between 10 and 16 percent protein, "and also a hard white winter wheat called Golden Spike that was developed a couple of years ago and has 12 or 13 percent protein." He sows his wheat in late August, and it takes eleven months before the crop is ready to harvest in the following July. That's because of the limited rainfall in Utah. He's tried improving his soil by plowing down cover crops of peas and using soil amendments with lots of micronutrients, but what works best for him "is chicken manure," a high-nitrogen manure that gets crops off to a healthy start. He's noticed that in some areas of his acreage, the amount of organic matter in his soil has increased, despite its being cropped each year. That's good organic farming and the very definition of sustainable.

He gets $5.60 to $5.70 per bushel of his organic red winter wheat, compared to the going price of $3.20 for regular wheat. "Your readers should understand why organic wheat is priced higher," he says. "A lot of people think organic farming is simply not using chemicals, but there's a lot more to it than that. There are inspections by certifiers that have to be paid for. Organic techniques take more insight into how the land and nature work. I have to plan ahead for insect control rather than just spray when the Russian wheat aphid shows up. This takes time—I have to be out in the fields, examining plants. I had to learn a different way of farming, such as changing planting dates to avoid insects and building the soil as I'm farming it. My tilling practices have to be timed to kill Russian and Canadian thistle (two noxious weeds) seedlings. Only then can I plant wheat. I also have to be a lot more careful than conventional farmers in how I store my wheat. All my sheds have aerators to keep down heat and moisture that stimulate weevil eggs to hatch. It also keeps down molds that could grow on the wheat.

"But there are other reasons why I'm organic. The first is economic. It's worked out financially for me to have different markets than the standard ones. Another is that I don't have to work with toxic chemicals. I think of my family's health and those who'll eat the products we grow," Grover says.

Chefs around the country, sparked originally by Alice Waters's dedication to organically grown foodstuffs, have formed the Chefs Collaborative and are demanding pure, wholesome organic foods, which creates the niche market that Grover has found to be profitable.

Once the grain is picked up by Giusto's, it goes to one of the distributor's own mills in California or Utah, where it waits until orders come in. The grain is blown pneumatically from place to place in order to remove dust and by-products before being milled. It's then milled to order, using roller, stone ground, or hammer mills with an air-cooled system that keeps the product as cool as possible to preserve vitamins and enzymes. Organic flours are then aged for two weeks to allow the natural enzymes in the wheat to reach full development. In the baking process, these enzymes turn starch into sugars capable of being fermented by the yeast. The flour is tested for protein and ash, then tested with a faringograph, a device that records the dough's resistance to kneading over about fifteen minutes or so, indicating its gluten development and toughness. The flour is then bagged and sent to buyers around the United States. (Incidentally, instead of a faringraph, the French use a device they call the *alvéographe de Chopin,* which tests a flour's elasticity by blowing a bubble of air into the dough and seeing at what point it bursts. The alveograph is named for the man who developed it around 1920, Marcel Chopin.

Suzanne Dunaway of Culver City, California, is a baker who loves organic flour but complains about its cost. Her Buona Forchetta Hand Made Breads was voted the best bakery in Los Angeles by *Los Angeles* magazine. "I started out all organic," she says, "but now I make so much bread that the cost of being all organic is prohibitive. Fifty pounds of organic flour is $14 compared to $7.50 for regular flour." Now about half her loaves are made from organic flour and the other half from conventional flour. Still, she says, "Organics put us on the map. Like my customers, I love knowing where my food comes from and knowing what they *don't* put in it."

She makes bread in a unique way, detailed in her book *No Need to Knead*. She starts by making a sponge of flour and water that sits overnight and begins to ferment. "I use sponges for everything. It's really a no-nonsense method. I call it the One-Stir Revolution [an allusion to *The One-Straw Revolution*, a classic book on organic rice farming in Japan, written by Masanobu Fukuoka and originally published in 1975]. Next day, I mix the sponge into flour and water to make a very wet, sticky dough. Then I just let the bread sit. I handle it like a newborn baby, infrequently, very gently. I fold it rather than push and shove it. The more you knead bread, the denser the texture." Her breads are full of holes, chewy, and tangy with a slight hint of rye to boost the flavor. "It has a good crust," she says, "but it doesn't puncture the roof of your mouth." I've found Suzanne's advice to be sound. I've made bread without kneading, but it's so full of large holes that it's not very good for sandwiches. So I've taken to kneading the bread to get a slightly denser texture. You'll find my recipe for homemade bread beginning on page 225.

Has she struggled to get her bakery off the ground? "Well," she says, "the pitfalls of business are the tough part. But I just put the bread out there to see who salutes." Obviously, many have saluted, but not all. "The public in the United States doesn't know bread. We have to change many prejudices about bread. I hear people say, 'Oh, I don't eat bread. It's so fattening.' That bugs me. It's only the quantity of bread that people eat that makes them fat—not the bread itself."

A Hotbed of Organic Baking

Several factors explain why the coastal region of Marin and Sonoma counties in Northern California is such a hotbed of organic baking and good bread. First is the longtime commitment to organic food in the region. Because of its cultural and climatic influences, the region is known as the Provence of America. The three-hundred-plus-day growing season allows for nearly year-round production of crops, especially fine organic salad ingredients and vegetables.

A second reason is that the availability of these foods has attracted great chefs like Alice Waters, John Ash, and Thomas Keller, among many others, who consider locally produced organic to be a mark of high quality. The proximity to the Bay Area cities of Berkeley, Oakland, San Francisco, and San Jose means chefs in these places can get high-quality organic produce rushed spanking-fresh to their restaurants. Naturally, where chefs demand fine organic produce, they'll also be looking for fine organic bread. Produce in this region is truly cosmopolitan—it includes local oysters from the cold-water bays, lamb and beef from the hills, milk and cheese from the dairy farms, and world-class wines from the vineyards. You'll notice that many of these products are fermented, and there's an active community of fermenters in Sonoma County.

A third reason the area has so many organic bakeries is because of the late Alan Scott, who built brick hearth ovens. From Mendocino in the north to the Bay Area in the south, he has seeded the region with wood-fired hearth ovens that turn out amazingly delicious bread.

Finally, the social milieu of the region includes a population that will demand and buy fine organic bread, supporting the bakers who elevate the quality of life in the area. Where the bread, cheese, and wine are good, life is good. Although this part of coastal California is blessed with the climate for producing fine food, citizens who want this kind of quality in their foodstuffs and in their lifestyles extends from the Atlantic to the Pacific in North America. A trip to Europe shows us that much of the quality of fermented foods we've come to enjoy in recent years has long been established there, including grassroots movements for small, organic, locally produced fermented foods that characterize various regions and help give them a distinct identity.

Globalization and mass-produced food is a top-down phenomenon. Marketing decisions are made in sealed air-conditioned offices in major cities. Ad campaigns are dreamed up in skyscrapers. Food is made from ingredients from a hundred places and loses all of its local character, to say nothing of its quality, in the processing. These foodstuffs are virtually the same in supermarkets everywhere. It's as easy to buy Kraft singles in Seattle as it is in New York. It all has a suffocating sameness to it.

The fermented food movement, on the other hand, is bottom-up. Nobody ever imposed an organic and ferment-minded sensibility on the people of this country. It arose because people wanted pure, wholesome food, both fermented and unfermented, of rich, defined flavor produced by their neighbors in an environmentally safe manner.

A Tour of Fine Organic Bakeries
North of San Francisco

I recently took a tour of some of the fine organic bakeries in Marin, Mendocino, and Sonoma counties.

The Brickery is the bakery at Café Beaujolais in the seaside town of Mendocino. It produces thirteen different breads and specialty baked goods, including one of my favorite breads, the amazing Austrian sunflower bread. This heavy, dense, and utterly delicious loaf contains white flour, sunflower seeds, cracked wheat, oats, barley, polenta, millet, buckwheat, flax seed, soy grits, sesame seeds, water, malt, sea salt, and yeast. A slice is a meal, and the flavor is like a walk in a sunny field full of ripe grains where you can pluck a seed head and chew on the fresh seeds. You need good teeth to chew it, but each chew bursts open another seed and adds to the complexity of the flavor of the bread. It's best with nothing on it, and toasting it lightly only improves the flavor.

I spoke with Tomas Fiore, who, with Tim Bottom and Gary Zachary, bakes the Brickery's breads in one of Alan Scott's ovens. The breads are served in the restaurant and sold in many retail shops in the area. Fiore says the bakery was a hit from the start. "When the smell of wood oven–baked bread traveled through the town, people were led here by their noses, and we were accepted right away.

"I use organic ingredients because the flavor excels," Fiore says. "There's a noticeable difference in taste. Chris Kump and Margaret Fox, the previous owners, were in Austria and found the sunflower bread in a little bakery there. They brought the recipe back, but it was in German. Two years later they returned to Austria, took the recipe with them, and had it translated. Now the Austrian sunflower bread is our most popular bread by far."

But there's another reason Fiore is organic. "It has more to do with ethics," he says. "I'm interested in good health and right livelihood. There's a dominant corporate paradigm that wants to bypass locally produced ingredients, and I'm against that. All around the world, bread is made from just three ingredients: flour, water, and salt. [Yeast and bacteria come free in the air.] All the many variations in bread come from the hands-on stuff: how long it sits, the proportions of ingredients, how long it rises and relaxes, and so on." It's these slight variations that make for the wonderful diversity of bread, but there's no diversity in bread mass-produced by large corporations.

From Mendocino, I drove down the coast to Point Reyes Station in Marin County, a picturesque western town perched at the head of Tomales Bay, an inlet off the Pacific known for its oysters—and as the breeding ground of the great white shark. My first stop was at the building housing Tomales Bay Foods. This establishment sells locally produced, organic items, including the cheeses of Cowgirl Creamery, as well as many of the world's great cheeses imported from Neal's Yard Dairy in England. Cowgirl's cheesemaker Sue Conley and her partner, Peggy Smith, who was a chef at Chez Panisse in Berkeley for seventeen years, use 100 percent organic milk from the nearby Straus dairy to make an array of fine cheeses, including the semisoft Mount Tam, possibly my favorite American cheese. Mount Tam has a triple cream's silkiness, along with a rich, milky flavor as beautiful and fresh as the stunning scenery around Tomales Bay and the Point Reyes National Recreation Area and seashore. After buying a sack of goodies, including two rounds of Mount Tam, I drifted over to the Bovine Bakery (you may guess from the name of the cheese company and bakery that the area is one of America's premier dairying regions), where I spoke with Shannon Stapel, the manager. She says, "There's a trend toward organic products. They're better—less processed. People want that." Her fifty to seventy-five loaves a day are baked in one of Alan Scott's brick ovens using Giusto's flour, but she says her best-selling products are morning buns and bear claws.

Not far away, on the western shore of Tomales Bay, is Inverness Park, where Debra Ruff bakes a country white loaf using a natural starter, ciabatta, and a whole wheat seeded loaf. She sums up her decision to bake organic bread from fermenting starter succinctly: "Better quality, tastes better, customers demand it."

AN IDYLLIC BAKERY IN AN IDYLLIC TOWN

From there I headed inland to the quaint little town of Freestone in Sonoma County—a picture-postcard village nestled in the rolling hills just east of the oceanside bluffs. There Jed Wallach runs the Wild Flour Bakery using one of Alan Scott's brick ovens. Two young women kneaded bread in the back of the bright, airy shop while a young man chatted them up. The counterperson was a woman with a rolled-edge hat. The pace was unhurried but steady, measured by the reggae playing on the radio. Referring to this pleasant scene, Wallach says, "I didn't want to bake alone. I wanted a party every day." And why did he decide to make organic bread from local yeast and bacteria?

"It's self-evident," he says, referring to the pretty little town in which he's located. "It's going back in time, as if it were a hundred years ago. Most small villages then had a single baker." This is a nice dream, but a hundred years ago Freestone was part of the rapacious destruction of the virgin redwood forests of the region. Nearby Guerneville was then called Stumptown for obvious reasons, and Occidental and Freestone were on a railroad line that hauled giant redwoods from the clear cuts as fast as the trees could be felled. Today, Freestone does look more like a Vermont village than a California logging town, and Wallach has taken advantage of that to build his dream bakery.

Wallach's first career was as a maker of architectural stained glass. His artisanship has been transferred to baking. "I've always wanted to keep things as simple as possible," he says. "The less machinery, the better. So using organic flour and natural starter is part of that. And it fits the community. So many people here are aware of the cost of nonorganic production." By that he means that the environmentally conscious citizens of Freestone, Sebastopol, and Occidental—towns relatively close together—understand that conventional agriculture depletes the land, encourages erosion, and damages local ecologies, and that the chief purpose of mass-production bread factories is to make profit, not great bread. A few years ago, Sebastopol was the first town in the nation to declare itself a "pesticide-free city."

"We make about eight hundred loaves a day, all sold from this shop," Wallach says. "I love the exchange with the customers and seeing their delight in our products. I always decline to make more bread than we do now because I like having time to spend serving the customers. I've worked and lived in Europe, in Paris, and in the Pyrenees. I saw how people run their lives—how customers were recognized on a first-name basis and valued. So the people who work here value our customers. Customers are smiling when they leave. The experience of buying the bread is as important as selling the bread.

"The value of organic is that it's a statement about the environment we live in. I'm not interested in making pretty food, but rather in food that will contribute to people's health—whether they are our customers here or growers in other places," he says.

THE ULTIMATE ARTISAN BAKERY

After Freestone, I headed south to Petaluma. On the way, I began thinking about the French Laundry over in Napa County. It's been named the best restaurant in America, and chef Thomas Keller has been named best chef by the James Beard Foundation for both the French Laundry and for Per Se, his Manhattan restaurant. I can vouch for the fact that dining there is extraordinary in every possible way. Its focus is decidedly organic, and the French Laundry serves bread from the Della Fattoria Bakery in Petaluma, Sonoma County. And therein lies a tale of astonishing success that's come to a strong, intelligent, warm, and wonderfully down-home woman named Kathleen Weber.

Kathleen and Ed Weber own fourteen acres in Petaluma, where Ed grew up. The land had been a chicken farm when he was young, but after his parents passed away, it was rented out to nearby farmers. "In the early 1990s, I was working in

Marin County [south of Petaluma] and I was driving home in a storm," Kathleen says. It was a difficult drive that set her to wishing that "there must be something we can do with those fourteen acres. So I talked to our agricultural commissioner and he suggested I grow red currants." Kathleen wisely decided not to take his advice (Red currants? Sheesh!) and settled on growing organic potatoes.

"I was selling our fancy organic potatoes to fine restaurants like the Sonoma Mission Inn," she says, "but I'd been baking bread at home forever. I thought maybe people could come here to pick their own potatoes and buy a loaf of bread. But then I went into Il Fornaio [a bakery in San Rafael] and freaked out when I saw all their beautiful breads. My body filled with adrenaline. I bought one of everything they sold! At that moment lightning struck, and I started making bread in earnest. I used the Italian method of making biga (a pre-ferment of flour and water) and started turning out Pugliese. I remember taking a big three-pound Pugliese to a party, and people said, 'Why don't you go into business?'"

That was in 1995. She started selling a few loaves at her manicurist's shop. "I kept trying new recipes," she says. And she was still selling potatoes to the Sonoma Mission Inn. Mark Vann, a talented young chef who ran the Sonoma Mission kitchen at the time, needed some bread for bruschetta (toasted Italian bread spread with garlic, basil, and tomatoes). "So I brought him what we thought was enough for two days. By the time I got back home, he was on the phone asking for more bread— he'd already run out. Mark's other bread supplier at the time was selling him olive bread with the pits still in the olives, and customers were complaining. He asked the supplier if the pits could be removed, and the supplier said no. So Mark asked me whether I could do production. I said sure—the way an actor says 'Sure' when the casting director asks if she can sing and dance. He said he wanted from eighty to one hundred loaves a day."

Her second account was the French Laundry. On a hunch, she took some samples to that restaurant. Later, Chef Keller called and said, "I like your bread." "I jumped for joy," she says. With those accounts, she soon picked up more, and now keeps her two Alan Scott ovens going round the clock with hired help, and the help of her son, her daughter, and her daughter-in-law. Her eight-month-old grandson helps by cheering everyone up. "I don't just bake bread in the ovens," she says. "I've had a pig roasting in each one, too. You can't imagine what a wonderful thing it is to have wood ovens going, ready at all times."

She now produces ten breads: campagne is a country white made from organic wheat flour, water, starter, and salt. Her levain is a blend of wheat flours with a little rye. She makes baguettes, wheat bread with currants (tiny dried raisins, not red currants) poached in brandy, Kalamata olive bread, polenta bread, pumpkin seed bread, roasted garlic with Vella dry Jack or aged Gouda cheese bread, seeded wheat bread, and a top seller, rosemary and Meyer lemon bread. All are leavened with starter made from the ambient yeast and bacteria that float in on the breeze.

"Bread is alchemy," she says. We walked to a table in her yard where her old dogs showed little interest in us. She brought out a bottle of good wine and a loaf of the rosemary and Meyer lemon bread. I went to the car and retrieved a round of that superb Mount Tam cheese from Cowgirl Creamery, and we had a mini-feast of organic bread, cheese, and wine as we talked. "The Meyer lemon and rosemary bread came together when I was driving somewhere with some rosemary bread and a bag of Meyer lemons in the car, and I smelled their aromas mingling together. The inspiration for the bread struck right then."

We cut a bit more of the Mount Tam, poured another glass of wine, and tasted that bread with its incomparable smell of herbal rosemary and bright, tangy lemon. "When all the planning and thinking is done," she says, "you just have to let go and have faith that things will work out. That's when inspiration can strike. If you've got everything all figured out, then there's no room for something new and creative in your thinking."

I described the bread that I'd been experimenting with, searching for the ultimate bread recipe for this book. "Sounds like you're doing everything right," she says. "We don't let our starter get sour. We refresh it twice a day. We do very little mixing of the dough, just gentle pulling and folding. Overmixing makes the bread dense and you lose the holes that keep the bread airy. Our doughs are pretty wet

A WORD ABOUT GLUTEN INTOLERANCE

Celiac disease is a genetic disorder that damages the small intestine when gluten, present in wheat and barley and some other grains, is ingested. The immune system identifies a peptide chain in the gluten as a foreign invader and attacks it. Gluten intolerance is not the same thing as celiac disease, and the small intestine is not damaged. However, it causes the same kind of unpleasant digestive symptoms, such as diarrhea, flatulence, abdominal pain, and malabsorption of nutrients.

The treatment of both disorders is the same: the complete removal of all gluten from the diet until the body heals itself. There is some indication that after healing, some gluten-containing products may be reintroduced to the diet if the intestinal ecosystem has been strengthened and the wheat products have been fermented—such as a sourdough fermentation for as long as a month.

when they go into the oven. The white doughs are now 75 percent water, the levain is 82 percent water, and the whole wheat is 90 percent water. After these doughs are mixed and risen, we shape them and let them rise again in the couche. Then into the oven."

Her ovens are fired with dry eucalyptus wood and get very hot—up to 700°F. Then the fire is raked out and when the temperatures fall to precisely the right level, the bread goes in.

"Baking bread is magic," she repeats. "It's a thrill every time a whitish dough goes in and real bread comes out." As for doing it all with organic ingredients, one look around her property told me why. Beautiful gardens contained cannas, morning glories, sage, rosemary, pomegranate bushes, cactus, yarrow, bamboo, mimosa, and much more, all surrounded by cool green grass.

"I wouldn't think of doing it any other way than organically," she says. "I've always grown a great garden. I read *Organic Gardening* magazine religiously back in the 1970s when it influenced so many people.

"You know, commercially grown wheat sends out shallow roots that only ab-sorb the chemical fertilizers given to it. Organic wheat roots have to go deep, where they get what they need to add richness and flavor to the grain. It makes better flour. Overprocessed flours from chemically grown grain are gray, not gold, warm, and beautiful like organic flour. As soon as that organic flour gets wet, it gives off a great smell.

"I don't take quality for granted. My goal is to make my bread the best it can be—and being organic is such a big part of that," she says.

We finished our little lawn lunch of organic bread, cheese, and wine, and I thanked her as she loaded me up with bread and some of the expensive sea salt from Brittany that she uses. As I got in the car, I noticed a big orange-and-white tomcat snoozing on my car's rear window shelf. "Is this your cat?" I yelled.

"Yes," she shouted back. "That's Logan. People have driven off with him more than once."

CHAPTER 9

Cheese

Various reasons have been advanced for the invention of agriculture. Early nomadic hunter-gatherers were supposed to have brought back grains of wild barley and emmer wheat to their campsites, where some fell to the earth. When they returned to their campsites months or a year later, they found an abundance of grain growing there. That's plausible, but I propose another idea. One day the nomads captured a wild cow that had recently freshened, or given birth. They tethered her to a stake, milked her (carefully—she was wild), then found that the milk soon turned to curdled cheese. The cheese was so delicious compared to their normal diet of roots and grubs that they decided to stay right where they were, right by that cow. And in fact, they decided to go catch a few more wild cows and make more cheese.

Fast-forward several thousand years to the time of Julius Caesar. In 54 BC, Caesar invaded the British Isles and found the Britons making Cheshire cheese.

Fast-forward to today. The British still make Cheshire cheese, although most of it is factory made and lacks the quality of real, tangy, artisan Cheshire. Here and there, though, the old hand-done methods have persisted or are even returning. The Appleby family's Abbey Farm wraps its raw cow's milk Cheshire in unwaxed cloth, maintaining that this produces a more honest Cheshire than those finished with waxed wrappings. Real farmstead Cheshire today is colored orange with annatto, a natural coloring agent derived from the seeds of the annatto tree (*Bixa orellana*), which is a native of tropical America but has spread by cultivation to many tropical areas around the world. It takes anywhere from two months to a half year to age this cheese properly, after which it becomes the dry, crumbly, savory cheese with a rustic flavor that was undoubtedly enjoyed by those Celtic yeomen before Caesar arrived.

Similarly, in villages across Europe and into Eurasia, cheeses have been made since the dawn of agriculture. Each village—even each farm—had its own environment where the cows, goats, or sheep grazed upon the local grasses and weeds. If this farm was situated on high, well-drained land, more of the tough, drought-resistant herbs like wild thyme and chamomile might be in the animals' diet; if it was along a wet bottomland, the milk animals had more juicy, crunchy plants like wild fennel and sedges. As any farmer who has grazed cows on pasture infested with wild onion in the spring knows, the garlicky flavor of wild onion goes straight into the milk, and so, to some degree, the mix of forage creates a unique flavor in the cheese. In mountainous areas, herds were often pastured higher up in the summer and closer to home in the winter, with different forage—and thus different milk and different cheese—at each season.

Because life tended to be confined to the farm and surrounding villages, the indigenous microorganisms developed a special ecological mix that also characterized the cheese they colonized. Cheesemaking temperatures varied according to the climate and time of year. Different farms curdled the cheese differently—some with the extract of calf's stomach called rennet, some with thistle flowers, or fumitory (*Fumaria officinalis*), some with little black snails, some with lemon juice,

some with stinging nettles, and some just let nature take its course and waited for the milk to curdle by the agency of whatever bacteria happened to be around.

Some farmers ate their clotted cheese fresh at the cottage table—hence cottage cheese. Others set it aside to find it hardening up and becoming intense and sharp, like Tyn Grug, an organic cheese from Wales. Others found it growing deliciously moldy and pasty, Tomme de Savoie from the mountains of eastern France. Others found (obviously to their delight) that the cheese softened and intensified in flavor as it ripened, as with Brie or Époisses de Bourgogne. And legend has it that about two thousand years ago, a certain shepherd in southwest France stashed his lunch of bread and cheese in a cave while he went off to dally with a neighbor maid, subsequently forgot about the lunch, and came back some weeks later to find the cheese covered with a blue mold that imparted a rich, piquant flavor. The cave, of course, was located near Roquefort, France, and they've been making blue cheese there ever since. We know that the shepherd story goes back some two thousand years, because Pliny mentioned Roquefort cheese in the first century AD.

And so the cheeses of the Old World of the nineteenth century and earlier were an enormously diverse and interesting mélange that developed through the ages in discrete, isolated areas, and were mostly locally made and consumed. In America, handmade farmstead cheeses were ubiquitous in this nation of farmers, right through the nineteenth century. It's hard to know at this remove whether many reached the distinction of the European cheeses that were brought to such great refinement by master cheesemakers who passed their knowledge along generation after generation. By the twentieth century, though, industrialization had pretty much gobbled up the making of cheese in America. We had our eponymous American cheese—a block of almost tasteless paste—and factory cheeses that took their names from popular European types: brick, cheddar, Muenster, Swiss, Parmesan, mozzarella, and a few others. Manufacturers used processing techniques and additives to aim for consistency, but they got uniformity, homogeneity, and monotony.

Worse, dairy farming in America became mechanized in many ways. Small family farms were forced out of business by the hundreds in the 1970s and 1980s as the economies of scale achieved by large agribusinesses rendered the small farm uncompetitive. Corporate agriculture, however, is not as healthy for the land, the animals, and the consumer as small, organic family farms. A small organic farmer—especially one whose family has invested several generations in the same piece of earth—knows the farm in an intimate way that no factory farmer can. He knows where the pheasant lays its eggs and avoids plowing there. The factory farmer not only plows there, but bulldozes out the hedgerows to increase the plantable acreage, destroying the incredible diversity of flora and fauna that develops in a hedgerow ecology. The family farmer knows where erosion occurs and plants trees there. He knows where the deer bed down, where the wild geese land, where the soil is too shallow for deep-rooted alfalfa. More important, a savvy organic farmer knows that the farm animals have natures that must be respected if they are to be healthy. The hen scratches in the horse bedding for fly eggs, keeping down the fly population as she improves the nutritional content of her eggs. In organic agriculture there's no place for mistreatment of farm animals, from the largest horse to the smallest rabbit. Organic farms are ecosystems because they are composed of interlocking groups of living creatures that support one another and create overall health. They are not simply economic engines or factories.

The Discovery of Peasant Cheese

From about 1950 on, and especially since 1980, interest in old, peasant cheeses—the kind made on small farmsteads—has blossomed. And none too soon, since from Norway to Morocco, from Wales to India, globalization has worked to standardize products across national boundaries. Not only that, globalization has worked to threaten small-production artisan cheeses through sheer publicity and popularity. Fine cheeses, like fine wines, are more and more appreciated around the world by those who can afford them. The pressure is on to step up production to satisfy the world markets. This encourages the small cheesemaker to adopt factory techniques and move away from the old, slow, hand-done methods that made such great cheese in the first place. And "peasant" cheeses by their nature can be made only in limited quantities. There's only so much of that good milk to be had. And only so many places where the ecology of the microbiota is such that great cheese results. Any attempt at mass production destroys the quality that makes the cheese worth seeking out in the first place. But as Isaac Newton taught us, for every action there's an equal and opposite reaction. The fact that more and more people around the world, and especially in America, are discovering these wondrous cheeses creates new markets for our own home-grown artisanal cheeses. The organic, back-to-the-land movement of the 1960s and 1970s is finally paying off. Now organic breads and wines are easy to find. Cheese has taken longer to penetrate the organic marketplace, but it, too, is booming. The number of American artisan cheesemakers jumped to 85 in 1999, hit 147 in 2001, and is now many hundreds strong. The best part is that like their European counterparts, most small-production, artisan cheesemakers are organic in approach and intent, even if they aren't always certified. They just do things the natural way.

The audience for fine cheese continues to grow. Artisanal, a restaurant in New York City, served, at last count, 188 different cheeses, has an entire cheese room, not just a cheese cart, and employs a full-time "cave master," as its master of cheese is called. The descriptions of the cheeses range from "aromatic" to "challenging" to "assertive, bordering on mean." It may be suddenly chic to order the moldiest, most offensive cheese on the list, but fads pass. What lasts is quality, and with cheese, many of the best were here long before our generation came along and will be here long after we're gone. It's timeless quality that we're focusing on, and quality starts with healthy pastures that produce healthy animals that give pure, flavorful, healthy milk that is colonized by friendly bacteria and molds that make extraordinarily delicious cheese.

The rBGH Problem

One of the chief concerns that organically minded consumers have about milk and dairy products like cheese is that much, if not most, of the cow's milk in the United States is from cows treated with genetically engineered bovine growth hormone (rBGH), also known as bovine somatotropin (rBST), to stimulate their milk production. Today's dairy cows injected with rBGH produce 2.5 times more milk than the cows of the 1950s—10 to 20 times the amount of milk needed to suckle a calf, putting added stress on the cow, shortening its life span, and increasing the incidence of mastitis, an infection of the udder that is then treated with antibiotics. The Animal Protection Institute states that 17 million shots of antibiotics are given to cattle every year for infections related to milk production and other diseases. Antibiotics can transfer to a cow's milk, and many scientists and medical professionals have been warning for years that as humans are exposed to a plenitude of antibiotics, new and antibiotic-resistant strains of disease organisms are formed and will continue to form. To say nothing about the deleterious effects on our intestinal flora from continued low doses of antibiotics.

Another danger to people drinking milk from rBGH-treated cows is that these hormones can get into the milk and create adverse effects in humans who consume them. In a recent letter to Kraft Foods/Philip Morris, the Sierra Club asked Betsy Holden, CEO of Kraft Foods, to stop the use of rBGH in all Kraft's dairy products. The letter states that "there is evidence that this genetically engineered hormone, given to increase milk production, also increases the production of IGF-1 in the milk, which has been shown to promote breast, prostate, and colorectal cancers." For this reason, wise buyers look for cheese that's either organic (rBGH is not allowed under USDA organic rules) or from dairies that certify they do not use this growth hormone. Many producers are proud of the fact that their milk is rBGH-free, and they advertise the fact. Sheep's milk and goat's milk are always free of the hormone, since it cannot be used on these animals, and Monsanto, the company that engineered rBGH, has not developed milk-stimulating hormones for them. Although Monsanto had enough clout to get rBGH's use permitted in the United

States, it is not allowed in Canada, France, Italy, Ireland, Great Britain, the Nether-lands, Belgium, Luxembourg, Spain, Portugal, Germany, Austria, Switzerland, Nor-way, Sweden, Finland, Denmark, Greece, New Zealand, or Australia. It seems that while our neighbor to the north and partners in Europe and the antipodes are con-tinuing to make cheese without the need to stimulate milk production by using dangerous hormones that may harm both the bovine and human populations, the United States continues to capitulate to agribusiness and its focus on profits before people. It doesn't have to be that way. Kraft makes cheese in Italy without the hor-mone because its use is banned there, so the company can't argue that it can't be done. It can only argue that rBGH makes factory cheese more profitable.

And that argument about profitability may not hold water for long. According to the polling and metrics service SPINS/ACNielsen, the organic dairy industry has experienced tremendous growth in almost every category it tracks. "I think these days American consumers understand the difference between organic and con-ventional milk," says Deirdre Fitzgerald, marketing manager for Stonyfield Farm, based in Londonderry, New Hampshire. "Consumers concerned with how the cows are treated and where their milk is coming from choose organic."

While big agribusinesses are driving small conventional family dairy farmers out of business (*Rural Cooperatives* estimates that the number of dairy farms in Wis-consin alone has decreased by 39 percent since 1989), organic dairying is one bright spot in family farming. Coulee Region Organic Produce Pool (CROPP), a Wisconsin cooperative made up of family farmers, has seen its membership increase from seven dairy farmers in 1988 to 1,600 from thirty-four states and Canada today. In California, shelves of large supermarket chains like Safeway, Ralphs, and Raley's routinely stock organic milk from producers like Alta Dena, Clover, and the Straus Family Creamery in Marshall, California. Besides making their own cheeses, Straus sells to local cheesemakers like the Cowgirl Creamery in Point Reyes, not far from Marshall. Cowgirl exclusively uses Straus organic milk and makes a variety of cheeses worth seeking out. Among them is quark—a tart, fresh, fermented cheese that will be known to those who've visited Denmark and Germany, where it's a favorite slathered on rich whole grain toast for breakfast or spooned over fresh fruit as you would yogurt. The company also makes crème fraîche, another fermented fresh cheese with the consistency of a loose yogurt, with a full buttery flavor and a sweet almond character. Crème fraîche is one of the secrets French cooks have for improving a wide variety of dishes by making them smooth, creamy, and fresh tasting. Cowgirl also makes Red Hawk, a sumptuous cheese defined by wild bacteria native to Point Reyes; in fact, the cheese couldn't be made anywhere else. It's aged four weeks and washed with a brine solution that tints the rind a sunset red-orange. Cowgirl also uses Straus organic milk to make Mount Tam, an aged, white mold-ripened, firm-textured, triple cream cheese for snacking or for the after-dinner cheese board. The Mount Tam slowly unfolds layer after layer of rich tangy flavor as it melts in the mouth, and is one of America's great artisanal cheeses. A third cheese, Chimney Rock, is mild and dipped in a sugary after-dinner wine, then dusted with ground mushrooms. It's named for a geological feature on the coast of the Point Reyes National Seashore.

Careful organic dairy farms like Straus take the preventive approach to disease control, building the health of the soil, and thus the health of the herd. "Our milking cows are not allowed to get antibiotics," says Vivien Straus, the creamery's marketing director. "In our case, we use homeopathic remedies."

Many organic dairies also don't homogenize their milk; that is, put it through a high-speed filter with tiny holes that break up the fat globules and disperse them through the milk so the cream doesn't float to the top. Years ago, most whole milk came in glass quart bottles that often had a bulbous top where the cream would rise and collect. Homogenization became standard practice in the late 1940s and early 1950s. And yet there is some question about homogenization and its health effects. In their *Extension Goat Handbook*, Fact Sheet E-1, G. F. W. Haenlein and R. Caccese of the University of Delaware state, "It appears that when fat globules are forcibly broken up by mechanical means, it allows an enzyme associated with milk fat, known as xanthine oxidase, to become free and penetrate the intestinal wall." Once xanthine oxidase gets through the intestinal wall and into the bloodstream, it may scar the heart and arteries, which in turn may stimulate the body to release cholesterol into the blood in an attempt to lay a protective fatty material on the scarred areas. This can lead to arteriosclerosis, they suggest.

This potential effect is not a problem with unhomogenized cow's milk. In unhomogenized milk, this enzyme is normally excreted from the body without much absorption. In a further e-mail exchange, Dr. Haenlein told me that "unpasteurized (raw) milk usually is not homogenized." He added that the final word is not yet in on the role of xanthine oxidase in cardiovascular disease. Whatever problems that may result will be avoided if your cheese is made from raw or unhomogenized organic milk.

The Raw Milk Debate

Just as a conventional farmer protects his crops by killing bugs with pesticides, killing the weeds with herbicides, and killing the soil microorganisms with fungicides, so also many people believe that sterilizing food will ward off disease. But we've learned from the farm model that far from warding off disease, sterilization eliminates the diverse ecologies that nature establishes to counteract disease organisms, and that contact with all kinds of microbes is necessary for the development of a healthy human immune system.

Lately there's been a lot of talk about prohibiting the sale of raw milk for making cheeses, both those produced domestically and those produced abroad. But almost all the world's great cheeses are made with raw milk, and until the late 1940s, almost all American cheese was made with it. Disease outbreaks were very rare because the lactic acid in the cheese produced by the starter cultures of lactobacilli microorganisms and the salt used to flavor and preserve cheese prevent harmful bacteria from growing. In 1949, a law was passed requiring that raw milk cheeses be aged for sixty days at 35°C (95°F)—enough time to kill off any harmful bacteria such as *E. coli*, salmonella, and listeria. This law was later extended to foreign-made cheeses, and over time, many states passed laws requiring pasteurization for cheeses aged less than sixty days. Still, the law didn't prohibit raw milk cheeses, and as the interest in good raw milk cheese both domestic and foreign burgeoned in the late 1990s, the agribusiness cheese industry began to take notice. In 1998, an industrial cheese trade group proposed that all cheeses made or sold in the United States should be made with pasteurized milk. The next year, the FDA informed the American Cheese Society, a group that represents artisanal cheesemakers, that it was investigating the survival of pathogens past the sixty-day waiting period, and that this investigation could lead to a requirement for pasteurization for all cheese. What a boon for huge corporate cheese factories if not only the artisan cheesemakers of America could be denied their superior raw material, but also pesky competitors like Parmigiano-Reggiano, Gruyère, Brie de Meaux, farmhouse cheddar from England, Roquefort, and additional scores of the world's greatest cheeses could be prevented from being sold in America. Those Europeans who weren't snickering at our naivete were horrified. They viewed any potential ban of raw milk cheese as both a witch hunt against microorganisms and a denial of their cultural heritage.

There was such an uproar that the FDA backed off its "ban raw milk" stance.

The fact is that raw milk cheese is simply better-tasting than pasteurized milk cheese—and safer. To quote one French study, in the *Journal of Dairy Science,* "Raw milk cheeses had a more intense flavor than the pasteurized milk cheeses. This difference was supported by the [chemical analysis] profiles of volatile [flavor] compounds. These differences were related to the high level of indigenous microflora in raw milk cheeses."

Pasteurization destroys the almost infinite possibilities for flavor and texture variations in cheese as produced by our friendly microbes. As with wine and bread, quality results from good, organic, raw materials and the proper manipulation of bacteria and yeast. Americans traditionally have felt safe with the sterilization approach, but due in no small measure to the organic community and its understanding of the beneficial role of microorganisms in health, the value of raw milk in the creation of world-class cheese has been recognized. The American Cheese Society and the Oldways Preservation Trust, a Boston-based nonprofit organization devoted to sustainable agriculture and traditional foods, joined forces to create the Cheese of Choice Coalition (CCC) to fight any laws banning raw milk cheeses. Speaking about the possibility of a ban, "We could become nothing more than a Velveeta nation," complained K. Dun Gifford, president of Oldways.

Slow Food, an Italy-based organization devoted to natural foods prepared by hand, has issued a "Manifesto in Defense of Raw-Milk Cheese." It says in part, "The bacterial health of our unpasteurized dairy products is destroyed by overzealous sterilization procedures. So will the health of human beings be destroyed through a diet of sterile food. Without any challenge, our immune system will fail and our medications become ineffective. Moreover, the unique flavor and aroma of cheese are conserved by non-pasteurization."

CCC has argued that some of the most often-quoted examples of disease outbreaks tied to cheese have been traced to fresh Mexican-style cheeses made in home kitchens, sometimes with pasteurized milk. Most of the recorded outbreaks of cheese-borne disease have been traced to pasteurized cheeses or resulted from improper production and unclean facilities. The CCC warns that pasteurized milk can be more vulnerable to pathogens than raw milk. The organic-minded can see why: The milk is deprived of its natural, healthy mix of microorganisms that produce both lactic acid and bacteriocins.

Part of the problem is the trend toward globalization and the worldwide standardization of food rules, which have the effect of eliminating the local, unique, and handmade in favor of the standard, factory-made processed foods sold by international food conglomerates like Kraft Foods/Philip Morris. This trend is characterized by the conglomerates as a way to protect human health, when in fact there's evidence that it may actually threaten human health. But what about legitimate health concerns regarding raw milk? Is there no danger at all?

Improper handling and unclean conditions are always a danger, whether in the small fromagerie or in the large factory, and no matter what type of milk is used. And in fact the danger may be greater in large cheese factories that import milk from many dairies over which they have little quality control. However, the Interñational Dairy Foods Association has promulgated a system called HACCP (Hazard Analysis and Critical Control Points); by keeping records and implementing HACCP properly, any dairy and cheesemaking operation can be kept clean and wholesome. It is the implementation of HACCP that the FDA should have been encouraging, not the prohibition of raw milk cheese.

Still, there was a time before pasteurization when tuberculosis, brucellosis, diphtheria, scarlet fever, and other deadly diseases were spread through poorly handled raw milk. Pasteurization was a huge step forward for public health after it was routinely introduced in the late nineteenth century. But with the advent of a better understanding of microbiology, better hygiene, modern milking techniques, and government regulation, raw milk, especially for cheese, is safer these days. Proponents of raw milk cheeses may also back the public's right to drink raw milk today. In fact, I use raw whole milk from a licensed dairy here in California when I make my daily kefir (the recipe is on pages 163–68), knowing that natural acidity in the milk, especially from the lactic acid produced by the kefir's lactobacilli, protects me. Most pathogens like a pH nearing neutral to alkaline, while the kefir is in the protective range of around pH 4.0. Besides, California health officials monitor that dairy like hawks or it would be shut down pronto.

The Source of Cheese Flavors

By what magic is milk changed into the sensual, rich, and delicious food called cheese? The magic of microbial activity. In the first decade of the twentieth century, L. A. Rogers of the USDA's Bureau of Animal Industry wrote a seminal paper entitled "The Relation of Bacteria to the Flavors of Cheddar Cheese" that pinpointed bacterial enzymes as the agents through which milk turns to cheese. Even when the bacteria are dead and gone, their enzymes remain in the curd, allowing for the continued ripening of cheese and flavor development for months and even years.

Today, we know it's not always bacteria alone, but bacteria and yeast together that enhance the flavor of ripening cheese, just as bacteria and yeast produce sourdough bread and kombucha. In fact, cows make the milk, cheesemakers coagulate the milk, but microbes make the cheese. Dr. Antonio de Almeida of the Catholic University of Portugal studied a raw sheep's milk cheese called Serra da Estrela and found six classes of bacteria (lactococci, lactobacilli, leuconostoc, enterococci, enterobacteriaceae, and staphylococci) plus yeasts. Analysis "revealed correlations between the major microbial groups present and patterns of volatiles generated." These "volatiles" are the flavor and aroma components that make cheese into a multisensory experience, with layers of complexity. The bacteria and yeast act on large, long-chain molecules in the milk and reduce them to simpler molecules that give good flavor to cheese. Sugars, proteins, fats, and glycerides in the milk are turned into volatile acetic, propionic, isobutyric, and isovaleric acids, plus semi-volatile fatty acids, free amino acids, and ethyl esters, the latter being responsible for fruity flavors in the cheese.

Yeasts join bacteria in producing the strong, pungent flavors and aromas of cheese. For example, with Limburger cheese, yeast first produces citric acid in the milk and then *Brevibacterium linens* converts the citric acid into the stinky sub-stances we associate with Limburger. Either bacteria or yeast alone will ripen cheese, producing characteristic flavors, but it's when the two team up that really interesting cheese results.

Food scientists have taken note of this and the genetic engineers have moved in. An article in the *Journal of Dairy Science* recently mentioned the "selective cloning and overexpression of the key enzymes involved" in cheesemaking. I investigated the term "overexpression" and found that it can mean a form of genetic modification. Lactic acid bacteria have a gene for manufacturing the enzymes that work on milk to produce flavor compounds in cheese. This gene has an "on" switch and an "off" switch. Genetic engineers find that switch and turn it on permanently in order for the gene to overexpress itself and keep producing enzymes even when they're not needed. Does this lead to better-tasting cheese?

A study done at the Department of Food Microbiology at University College in Cork, Ireland, suggests not. The scientists took cheese bacteria and modified them genetically in various ways, turning on this gene, adding genes from other bacteria, and so on. They also made cheese using plain old natural *Lactococcus lactis* subsp. *lactis* (one of the main bacterial starters used in cheesemaking) as a control. The results? "Taste and chemical analysis showed that the control cheeses were of the highest quality." Well, yes.

Of course, it's not flavor the genetic manipulators are after—it's consistency and cost economy. The summary of the conference notes from a recent seminar in France organized by Systems Bio-Industries states that "developing standard cheeses requires control of the diversity of microflora."

Large cheesemaking operations seek product uniformity, but the diversity of microflora at work in curdled milk causes enormous variations in flavor and aroma. And so industrial cheesemakers remove all microorganisms through pasteurization, then add starter cultures to achieve uniformity. That's great if simplified flavor is what you're after. The seminar's summary complains that the difficulty in identifying which bacteria produce which flavor compounds limits the ability to develop improved starter cultures. It suggests that electronic "noses" might assist the process. So, not only is nature's own mix of microbes far too unwieldy and unpredictable for the industrial cheese gang, but the progress toward standardization will be augmented by electronic noses. Perhaps the end product should be fed to robots instead of people.

The best cheese flavors result from bacteria and yeasts working together on the sugars, fats, and proteins in milk. Some strains produce enzymes that convert milk sugar (lactose) to lactic acid. Others convert milk fat to flavorful free fatty acids. Others change proteins into their constituent free amino acids. Some of the most interesting flavors in cheese arise when strains of bacteria produce amino-acid-converting enzymes. These convert the free amino acids, especially methionine, into flavor-active sulfur compounds. Such complex actions happen on the molecular level, far too small to be seen by the naked eye. However, we can taste the results in the best organic farmhouse cheeses. Natural and organic cheesemakers treasure the microbes that make their cheese.

If you plan on making some of your own cheese, this material on milk should interest you. You want what nature, through her profligate microbial fecundity, can give you—unique cheese made in your own kitchen. The quest starts with the best raw material: organic raw milk. And if that's unavailable to you, then organic pasteurized milk will do. Although it's sterilized, you can recharge it with a culture of mixed bacteria and yeast put together expressly to make cheese; culturesforhealth .com sells kits for making just about any kind of cheese you desire. Or you can let the milk sit out and curdle naturally, drain off the whey and refrigerate it for other purposes, press the curds (or leave them as is for cottage cheese), and see what you get. That's the way it has been done for thousands of years.

A Visit to an Artisan Cheesemaker

In 1986, Cindy and Ed Callahan moved from San Francisco, sixty miles north to Sonoma County, California, and the grassy hills along the Pacific Ocean. The water is too cold for swimming here, owing to the Japanese current that swings up the eastern side of Asia to Alaska, then down the western side of North America. It does produce a strong temperature differential between the cold water and the warm land that creates strong afternoon flows of onshore cold air and fog. The cool temperatures and grassy hills make it perfect sheep country, and so the Callahans began raising sheep on their thirty-five-acre farm, along with one hundred acres leased from a neighbor. Lamb was their business, and Callahan's lamb soon became the standard for choice local lamb in the many fine restaurants of the wine country of Sonoma and Napa counties, and the San Francisco Bay Area to the south. After the lambs were born, they ran in the pastures with their mothers for thirty days, then were sold as very small premium lamb. The herd grew to four hundred animals—but after the lambs were sold, there was still a lot of milk. Wanting to expand the business, Cindy decided to try making cheese. She called her local county agricultural agent and asked him how to milk a sheep. "Milk a sheep?" came the incredulous reply. After trips to Europe and a perusal of the literature, and with the help of a Scottish woman who had worked at a sheep dairy in Scotland, Cindy, her sons Liam and Brett, and Liam's wife, Diana, began California's first sheep dairy in 1990.

Bellwether Farms cheeses, as her products are called, are named for the lead sheep ("wether," originally a castrated ram) in any group, on which the farmer hangs a bell so other sheep can hear it and follow. Hence, that lead sheep is the bellwether. Bellwether's cheeses rank among the most sought-after farmstead cheeses in the country. They are extraordinary. I first tried one—the Toscano, a Tuscan-style hard cheese about the consistency of a firm cheddar—in the early 1990s and had never before tasted a cheese that was so pure, buttery, delightfully nutty, and rich. Although the farm isn't certified organic, the only difference between Cindy's farm and an organic operation is that the feed she uses when pasture is unavailable is not organically grown. Everything else is as down-home and natural as can be. I spent a morning at Bellwether with Liam and Cindy, and they took me through the whole process. Now I'd like to take you through the process, too. Whether your interest is in making cheese, using it in recipes, or just enjoying it with a glass of wine and a crust of good bread, your knowledge and enjoyment will be enhanced by this peek into a real artisan cheesemaking operation.

I arrived at about seven when the morning milking was just getting under way. Liam, a large, stocky man of thirty-five, wearing alligator green and cream-colored Wellingtons, an off-white full apron, and a baseball cap with the Bellwether Farms logo, oversaw the operation. The ewes were brought in twenty-one at a time to fill the milking stanchions. A trough with cracked corn kept them occupied while the milking proceeded. The first ewes in didn't walk all the way to the end of the line to start nibbling the corn, and so they were bumped and nudged along the stanchions by the rest of the group until all stations were filled. A sheep's udder fills with milk into two defined football-shaped compartments joined together along one edge, with one teat angled slightly outward on each. With the ewes happily crunching corn, the milking was carried out by two workers who slipped the milking machine's cups over each of the ewe's two teats.

The sheep were mostly East Friesians, the northern European breed of preference. This breed gives about five pounds of milk a day from the morning and evening milking. The top 1 percent give eight-plus pounds. This is a lot for a sheep. According to Liam, the milk of Friesians has fewer solids but makes up for it in volume and in how long the sheep can be milked before they dry up. Mature ewes from three to ten years old will be milked for eight months each year. Then they'll be impregnated and relax until lambing time in February to May. That freshens them for another eight months of milking. Also in the stanchions were Dorsets and a scattering of other breeds that give about two-plus pounds of milk a day—less milk but of a richer consistency, with more solids.

These sheep have fresh, green pasture grass until June, when the annual summer dry spell turns the hills their famous golden color. Green doesn't return until the winter rains set in, usually in November. During the dry season, the sheep are fed supplemental conventionally grown alfalfa. Since cheesemaking has become the center of the farm operation, the herd is down to about 175 animals. Demand for Bellwether sheep's milk cheese has become so great that the Callahans diversified by buying milk from a neighbor's herd of Jersey cows and began making a selection of cow's milk cheeses that have become as popular as their sheep cheese. "That took the pressure off this farm," Liam says.

I noticed that the workers would finish milking a ewe, then return to her later for another go-round. "Sheep often have a second let-down, as it's called, when more milk appears after the first milking," Liam said. "It's usually richer, and it's

good to milk the ewes out completely." A few of the ewes who had just lambed had colostrum (in sheep, the thick, yellowish fluid nature provides initially to give young mammals a good start), but their lambs had died, and so their colostrum was milked out by hand, frozen, then fed to any weak newborn lambs to bring them to good health.

Milking dilates the teat canal, so the workers wash the

teats with a combination of iodine solution and emollient to protect against infection and aid healing. At first, Liam said, the farm had problems with mastitis, which is usually cured by antibiotics on nonorganic farms. But on their visits to sheep farms in Tuscany and on Sardinia, they noticed that at the bottom of the teat cups the farmers had made a small hole in the evacuation tube that carries the milk from the animal to the tank. When they returned to California they punched similar holes in their tubes. Liam showed me how, when he held his finger over the holes he'd punched in his evacuation tubes, the suction surges of the milking machine would produce little back spurts that drove milk back into the teats. That was causing most of the mastitis, and the problem was solved by the simple measure of puncturing the suction tube so air from the hole would prevent the back-surge from reentering the teat.

The milk flowed from the cups to a pipe that carried it to a tank where it was cooled to 35°F. The milk was kept in the tank until there was six days' accumulation (later, in the high milk season in summer, it would be three days' accumulation)—about 120 to 140 gallons. Now, I grew up next to a dairy farm, so I asked the obvious question: "What happens when you have a power outage and the milking machines don't work?" The answer: "We milk a hundred and twenty-five ewes by hand twice a day." I whistled a long "Whew!" But Liam said, "We visited a farm near Siena, Italy, where four hundred ewes were milked twice a day by hand." I got uncomfortable just thinking about that, but Liam said the Siena farm was within a year of acquiring modern dairy equipment. "Most dairies in Italy have better equipment than we do, due to government support," he said. The first group of sheep was now finished being milked, and the ewes ran out as another group filed in for their corn, their udders swinging fat and full behind and beneath them.

Before we went to the cheesemaking room, Liam gave me a taste of fresh, raw sheep's milk. It was a beautiful light cream color, shiny and slightly viscous rather than pure white and watery with a slight bluish tinge like cow's milk. It tasted smooth and creamy, too—delicious! "Sheep's milk is like goat's milk—it's naturally homogenized. And it's rich, about 6 percent butterfat and 5 percent protein, compared to a Holstein cow's 4 percent butterfat and 3 percent protein—so the sheep's milk shows that richness in the mouth," Liam said. Bellwether's cow's milk is all from their neighbor's Jersey herd, and Jersey milk is 5 percent fat and 4 percent protein, which, like sheep's milk, makes a rich, flavorful cheese.

As he filled fourteen ten-gallon cans with the proceeds of six days' milking, then washed everything down, Liam said that people are looking for raw milk cheese now: "It has a greater perceived value." He talked about cheese as milk preservation. "It's a traditional way of extending the time you can use milk. You can dry milk to a powder. You can sour it, and you can salt it. The souring with bacteria is particularly important. In raw milk, the balanced blend of natural bacteria prevents growth of strains that produce toxins." From my research, I realized he was talking not only about lactobacilli turning lactose into lactic acid, but also about bacteriocins, compounds from beneficial bacteria that inhibit pathogenic bacteria. "The FDA is looking at one bacteriocin for its possible medical use," he said. He wasn't sanguine about the future of raw milk cheese, though. "It's an inspector's job to be proactive. When you have someone who's employed to look out for the public safety, once they solve one problem, they have to go on to the next, which leads to zero tolerance for any possible danger, whether it's warranted or not," he said.

The cheesemaking room at Bellwether was once a room where young calves were kept in tight confinement, fed milk, and raised only long enough for them to produce the white, milk-fed veal prized by some Italian restaurants. The practice is increasingly recognized as cruel. Most of today's veal is raised on pasture or in more humane conditions in barns. Certainly cheesemaking is an infinitely more benign use for this room, especially after seeing those happy ewes eager to file in for their morning corn and milking.

The room contains all stainless-steel equipment, bright overhead lights, and thick plastic sheeting on the walls so everything can be hosed down and cleaned after use. The equipment includes a double sink, stainless-steel tables, a stainless-steel pasteurizer with a digital readout (all Bellwether's cow's milk cheese is pasteurized), a tank with a hot water jacket used to warm the milk before turning it into cheese, and a 160-gallon-capacity tank where the sheep's milk is curdled. This large tank sits up on a grillwork catwalk and has a lever that allows the cheesemaker to lower the front end so all the curds and whey will run down toward a valve inserted at the bottom of the front end and the tank can be completely emptied. As Liam's workers poured the ten-gallon cans into the warming tank, I asked Liam where he'd learned how to make cheese.

"I took short courses at Cal Poly," he said. "Knowing the science helps you understand the process." He said that there are a lot of factors that influence the texture, aroma, and flavor of the finished cheese. "Most of the flavor in cheese comes from the enzymatic breakdown of protein. First the bacteria convert lactose to lactic acid, then they go for the protein. The fat takes longer to break down, but it is eventually converted, too" (to short-chain fatty acids; some scientists contend that it is these fatty acids that contribute the majority of the flavor of a cheese, but others agree with Liam that it's the conversion of protein to free amino acids that creates the yummy flavors we love so much). "The cheese flavors are dependent on what kind of starter bacteria you use, the type of rennet, the amount and kind of proteins in the milk, the type of sheep, the temperature of the milk, the moisture content of the cheese, the temperature of the ripening room, the strains of wild bacteria and yeasts that are in the milk, and the natural flora that's on the equipment and around the dairy—it's very complex and there's no way to know all the variables. But we try for consistency," Liam said.

First the 140 gallons of sheep's milk went into the warming tank, where it was warmed to about 100°F. While this was happening, Liam sanitized the cheesemaking tank with a chlorine and water solution. "Before electricity, people used to warm milk over a fire. They still do that in places like Sicily. Or they use the milk warm from the animal."

Cindy Callahan came into the room to help with the actual cheesemaking. She'd been a nurse and then a lawyer in San Francisco, where her late husband, Ed, had been a physician. What started in 1990 as a little local cheesemaking operation soon became well known in the wine country. At first Bellwether cheeses were sold at farmers' markets, and the Callahans got lots of enthusiastic, positive feedback about their cheese. "Now we've got a tiger by the tail with this business," Cindy said. "We've filled every inch of this building. We've invested a lot of money in equipment. Do you realize how much cheese you have to make to support all this?" Despite her protestations, she's obviously proud of Bellwether's success.

The sheep's milk was now pumped from the warming tank into the large tank up on the grillwork. "Now we add the starter culture," Liam said, and tore open a foil package and shook the contents—a freeze-dried, cheese-colored powder that smelled like toast (Liam says like potato chips)—onto the surface of the glistening, creamy milk. It reminded me of the Vermeer painting of the kitchen maid pouring thick, creamy milk from a crockery pitcher into a bowl. The culture is a mesophilic culture called MM 100. It's made in France and sold by the Dairy Connection, a supplier of cheesemaking cultures based in Middleton, Wisconsin. It contains three strains of lactococcus bacteria—*Lactococcus lactis* subsp. *lactis*, *Lactococcus lactis* subsp. *cremoris*, and *Lactococcus lactis* biovar *diacetylactis*. The first two are the main lactic acid–producing bacteria used by the cheese industry, whereas the latter strain adds aroma to the finished cheese. As Liam stirred the starter culture into the milk, he said that he keeps the temperature of the milk in the upper nineties, near one hundred, as it's about to be made into cheese. "This is the upper range of the starter culture's comfort," he said. "Most cheesemakers who use the mesophilic culture keep their milk in the high eighties or lower nineties. I saw them doing it in Italy at the higher temperature, and it worked out fine. It's worked out for us, too."

We now had to wait twenty minutes before the coagulant would be added. Liam and the workers used the time to clean and wash the warming tank and hoses and wash down the floor.

Bellwether uses chymosin, the same enzyme as the rennet in calves' stomachs. But actual calf rennet is not legally sold in the United States, and so the analog chymosin, laboratory-made, is used. Liam took a pipette and mixed exactly 70cc of pure chymosin in a quart of distilled water. "This is a powerful enzyme," Liam said.

"If I'd put it straight into the milk, it would immediately form a hard clump, so I have to dilute it. They make this enzyme in a laboratory, but I think the natural is better. In Sardinia, I saw them making the traditional cheese called Fiore Sardo. There was a bucket of salt on the farmhouse floor with calves' stomachs in it. It had a distinct smell—not unpleasant but unique. Even though the cheese was aged and smoked for a month, I could taste in the cheese what I'd smelled in that bucket. Here in the States adding the coagulant is just a production step, but elsewhere it's crucial to the flavor of the cheese. I don't mind using chymosin, though. I think the milk flavors come through better." He poured the diluted enzyme into the milk and quickly stirred it through the vat. "You can vary the amount of enzyme—more gives you a firmer curd, less gives a softer curd. But all three of our sheep's milk cheeses will be made from this same vat. The difference comes from the ripening process and with our Pepato, the addition of peppercorns."

We now had to wait for thirty minutes as the coagulant worked. I asked Liam what he does with the whey—the liquid that separates from the curd and is drained off. "We use that for ricotta," he said. "We make ricotta from both sheep's and cow's milk whey—they're very similar, although there's a slight flavor difference. Ricotta was invented only a hundred years ago, you know. In Italy, they don't make cow's milk ricotta. They feel sheep's milk is superior." In Italy, whether whey goes to ricotta or not depends on the traditions of the place. Around Parma, Italy, for instance, whey from the making of the famous Parmigiano-Reggiano, one of the world's great cheeses, is fed to the pigs that yield Parma hams and prosciutto, another of the world's great foods.

"Ricotta isn't hard to make," Liam says. "We raise the temperature of the whey to 180 degrees, add salt, and acidify it with vinegar, then the solids precipitate out. We drain the solids and we've got ricotta." *Ricotta*, by the way, is Italian for "re-cooked," which is how the cheese is made.

While we were waiting for the chymosin to work, Liam and Cindy showed me the cold ripening room. Ripening rooms are used in lieu of caves in the earth, which naturally have the right conditions of temperature and humidity and are often used in places where cheesemaking is an ancient art and the caves were long ago devoted to ripening the local cheeses. The air in this ripening room was moist, the temperature was about 45 or 50°F, and the wooden slat shelves were filled with round, inviting-looking, ripening cheeses. The youngest were creamy white. "These are our San Andreas and Pepatos," Liam said. The San Andreas, named after the famous fault that caused the San Francisco quake of 1906 and which runs pretty much right under Bellwether Farms, is a smooth, full-flavored sheep's milk cheese aged for two months. The Pepato is a similar cheese but with whole black peppercorns added. The oldest cheeses in the room, aged there for four months, were a dark gold. These were extra-aged Carmodys, named for the lane on which the farm is located, made from Jersey cow's milk, with a smooth texture and a wonderful buttery flavor that Liam says is his take on the northern Italian Tomme. Much of the batch of milk that Liam was working with that day would be made into Toscano—a traditional hard Italian-

"That's called a clean break."

style cheese aged four months, with nutty and fruity overtones to its strong buttery taste. In 1996, the American Cheese Society's annual judging named Bellwether Toscano the Best Aged Sheep's Milk Cheese in the country, and one taste tells you why.

After thirty minutes, Liam and I climbed up on the grillwork and he patted the surface of the milk with the flat of his hand. It appeared rubbery, as though it had solidified. He took two fingers and slid them into the curd, then drew them up and out. The curd parted easily and cleanly and clear whey filled the slot he'd opened. "That's called a clean break," he said. "That's what you look for. If the cheese isn't set, the whey will still be milky." He showed me another way to tell whether the curd is properly set. "Some cheesemakers do this." He laid his hand, palm down and flat, on top of the curd where it met the vertical metal side of the vat. Exerting just a little sideways pressure, he moved his hand away from the metal side, gently pulling the curd from the metal. Again it came away cleanly, with clear whey in its wake.

"Cheese is ready," he said to a worker, and the worker handed up the cheese harp. This is a stainless-steel frame about three feet square—just the size of the vat— with seventy-two wires strung across it. Each wire is about a half inch from its neighbor, and there are three thicker wires that separate the thin wires into four sections of eighteen wires each. With Liam on one side of the vat and the worker on the other, they each grasped a side of the harp and lowered it vertically into one end of the vat, with the wires in latitudinal (horizontal) positions across the vat. They quickly drew the harp, holding it vertical, toward the other end, slicing the entire curd into half-inch horizontal slabs. They took the harp out of the vat and turned it 90 degrees so the wires were vertical, then drew it down the length of the vat in the same direction as the first pass. They lowered it vertically along one side of the vat and drew it across the curd to the other side, then reversed direction and drew it back toward the first side. They repeated this side-to-side action once more, then finished cutting the curd by drawing the harp from the far end of the vat back toward the front end where they first began cutting the curd. After the first horizontal cut, all the rest were done with the wires held vertically.

Then Liam and his worker did something that surprised me. They lowered their arms into the vat up to the elbows and began waving their arms and swirling them through the vat. "We have to mix by hand like this or the curd will clump back together," Liam said. "This controls the final moisture of the cheese, because it helps the whey separate out from the curd." They continued doing this for almost five minutes. Toward the end of this process, Liam said, "I can really feel the curds firming up—I can feel distinct bits of curd hitting my hands."

After five minutes, they stopped and quickly got down off the grillwork to stand at the stainless-steel tables under the valve at the bottom of the vat. The tables were filled with round drain baskets about eight inches in diameter, perforated with many small holes and rounded on the bottom. A rectangular plastic form was placed over six of these baskets to hold them steady. Liam opened the valve on the vat, and curds and whey shot out into a large bucket. When the bucket was filled, the worker took it away and Cindy handed Liam another bucket for filling. The worker began pouring curds and whey into the drain baskets in the form, filling each basket and moving to the next. After he'd filled six baskets, another worker moved the form to the next six empty baskets. When the worker emptied his large bucket, Liam had another filled one waiting for him. And so the process continued until most of the drain baskets were filled with curds to within an inch of the top. At this point Liam grabbed a large container of black peppercorns and added a measured amount to a large bucket full of curds and whey, then mixed them up and began filling the rest of the drain baskets, which would become the Pepatos.

Meanwhile, the whey was draining furiously from the baskets onto the stainless-steel table, which had a slight tilt toward the far end. The whey, a yellowish gold color with a slight tinge of green, ran down into a drain and then through a hose into large plastic containers.

When the vat of curds and whey was emptied, fifty-four filled drain baskets sat on the table and eighty-five gallons of whey rested at the far end. Then the workers began to flip the baskets. Using quick, circular back-and-forth motions, they flipped the curds—which at this point looked like cottage cheese—inside the baskets to encourage more whey to run out. Liam said the baskets would be flipped like this every fifteen minutes for two hours, after which the curd would have condensed to form a nice, softly rounded fresh cheese. The cheeses would then be turned upside down and the baskets removed. No pressing is involved in making these cheeses. "Pressing would make them chalky," Liam said.

To make Toscano, the cheeses are placed in a warm, 80°F room for twenty-four hours, after which they go into the cold, humid ripening room. The San Andreas and Pepato cheeses go immediately into the ripening room. Because these cheeses are natural-rind cheeses, nothing further is done to the rinds.

On Pairing Cheese and Wine

Cheese is milk's leap toward immortality.
—CLIFTON FADIMAN

When pairing wine with cheese, the best rule is to follow your own taste buds, but don't let either overwhelm the other. So a rich, inky Zinfandel may swamp the flavors of fresh cheeses like a fromage blanc, a ricotta, or a farmer's cheese. Similarly, very aggressively aromatic cheeses like Esrom and Limburger, very intense and sharp cheeses like aged cheddar and Swiss Appenzeller, and ripe soft cheeses like Italian Paglia and Toma may overpower delicate wines like a German Mosel. But then again, if you like your Limburger with Mosel . . .

The flavor of cheese depends in part on the milk used. In general, cheeses made from cow's milk have the mildest flavors, while goat's and sheep's milk cheeses have more robust flavor. However, ripeness also determines the degree of flavor. An aged cow's milk cheese will be stronger than young, fresh cheese made from either goat's or sheep's milk, and it's the bacteria and yeast that turn fresh milk into ripe cheese bursting with flavor.

Wines also range from very light, such as Sauvignon Blanc, Pinot Grigio, Pinot Blanc, Chenin Blanc, Sémillon, and Riesling; through medium-bodied and medium-flavored like fruity Chardonnay, Pinot Noir, Grenache, Counoise, and roses; to full-bodied and rich, like Cabernet Sauvignon, Zinfandel, Merlot, Syrah, Sangiovese, Mourvèdre, and some of the more concentrated Pinot Noirs. In a separate category are the sweet dessert wines: late harvest, botrytized, and fortified wines like ports, sherries, and sauternes. In Europe, wines are usually labeled by type or estate rather than grape variety.

While I certainly encourage you to try any combination that strikes your fancy, I find that matching light wine with light cheese, medium wine with medium cheese, and rich wine with intense cheese works well—however, the best overall rule of thumb is that cheeses go better with white wines than with reds. The very intense cheeses, like the blues, and sharp, nutty, salty, hard grating cheeses, such as Asiago, Pecorino Romano, and Parmigiano-Reggiano, pair best with the sweet dessert wines, including port.

Cheeses run from fresh cheese to soft-ripened, bloomy cheese (Brie and Camembert); semi-soft washed-rind cheese like Époisses, Taleggio, and Spanish Mahon; semi-soft cheese like the Dutch Gouda and Edam, Cantal, Cheshire, Fontina, Appenzeller, and Port Salut; hard cheese like Manchego, dry Jack, provolone, Swiss cheeses, and cheddar; and blue, such as Maytag, Cabrales, Roquefort, Gorgonzola, and Stilton.

And sparkling wine? Best with any food, best with any cheese.

A Few Wine and Cheese Pairings

I would love to sit down in front of a crackling fire with some Tuma dla Paja, some cognà, and a wonderful, wood-aged barbera like Monti's Barbera d'Alba. I would have this with an Italian rosetta roll, typical from Trieste. The creamy, moldy complexity of the Tuma is offset by the richness and sweetness of the cognà, and it goes perfectly with the rosetta bread, which puffs up like a rose and remains empty on the inside but pure and crunchy on the outside. The deep, berry-like qualities of the barbera wash all of it down—a delicious combination.

(Note: Tuma dla Paja is a superb Piemontese cow's milk cheese that comes tied up with straw. Cognà is a mixture of fresh and dried fruits stewed in wine for six to twelve hours.)

—LIDIA MATTICCHIO BASTIANICH, TV cooking show host and chef-owner of Felidia, Del Posto, Esca, and Becco restaurants, New York

Great Hill blue cheese from Marion, Massachusetts, drizzled with chestnut honey, with candied and ground chestnuts on the plate, served with walnut-date bread and a 1971 Climens Barsac. —DANIEL PATTERSON, chef-owner of Coi, San Francisco

Our Mont St. Francis aged raw milk goat cheese sliced very thin with a pain au levain and a Ridge Zinfandel. —JUDY SCHAD, cheesemaker, Capriole Goat Cheese, Greenville, Indiana

Our blue mold rind aged goat cheese with a crusty French baguette and a Châteauneuf-du-Pape.

—GREG SAVA, cheesemaker, Brier Run Farm, Birch River, West Virginia

Grafton Village Classic Reserve Extra Sharp Vermont cheddar with a crusty, Eastern European–style rye bread, like they sell at Orwasher's Bakery on 78th Street in New York City, with . . . hmm, a big cheese needs a big wine . . . a big Cabernet Sauvignon.

—PETER MOHN, vice president, Grafton Village Cheese Company, Grafton, Vermont

My Bear Flag Brand Dry Jack with bread from the Basque Boulangerie here in Sonoma when it's fresh from the oven, with Chateau St. Jean's Orange Muscat.

—IG VELLA, Vella Cheese Company, Sonoma, California

When I was a kid, my grandmother would serve Teleme, and she also had crusty Tuscan bread made without salt. I love them with a glass of Cabernet Sauvignon.

—TIM MONDAVI, winemaker-owner, Continuum Estate, St. Helena, California

CHAPTER 10

Wine

When I moved to the wine country in 1985, it was by pure chance that I bought a house a few hundred yards from the vineyards of Joe Swan, one of the great pioneers of the reemergence of fine winemaking in California, and the man who brought Pinot Noir to the Russian River Valley of Sonoma County. I was fortunate to know Joe for a few years before he died, and to taste some of his wines. One in particular struck me as a mighty drink—dark as the skin of ripe Bing cherries, spicy and rich as a crème brûlée, with fruit-forward flavors of blueberries, dusty blackberries, and plums. It was a bottle of Joseph Swan Vineyards 1970 Gamay, at that time fifteen years old, that Joe had made from fruit grown in the Dry Creek Valley fifteen miles north of his property. I never forgot that glass of wine, but it struck me as curious that a Gamay—usually a light, fruity, inconsequential wine—could be so powerful and massive.

And herein lies a tale: A number of years ago, I dropped in on a party where the host was pouring a thirty-one-year-old bottle of, yep, 1970 Joseph Swan Gamay (probably the last bottle in existence, I thought). Amazingly, it was still as young and fresh-tasting as I remembered from sixteen years before, just as jammy and flavor-packed, but now silkier with age, and with an aroma that had only improved with time. I had discovered along the way that the grape variety wasn't Gamay at all; it was Napa Gamay, a name that had been given long ago to a mystery grape that no one could recognize. In the late 1980s, however, the great French ampelographer (a scientist who identifies grape varieties) Pierre Galet visited the United States, and on his tour of California identified Napa Gamay as a French variety called Valdiguié. Amazingly, the very next night I attended a dinner party where one of the guests brought, yep again, a bottle of 1970 Swan Gamay. He claimed to have several more bottles stashed in his wine cellar. He also told me that Andy Cutter of Duxoup Winery in Sonoma's Dry Creek Valley has been making wine for years from the same Valdiguié vineyard where Joe Swan bought his fruit thirty-one years ago. I called Andy to see if he was still making the wine, and he reported rather sadly that the vineyard owner ripped out the vines after the 2000 vintage, replanting it with Syrah and Sangiovese, two varieties more in favor these days.

Andy said that thirty years ago, Valdiguié was bringing a premium price, because Gallo, the huge winery in Modesto, used to purchase the variety to add to its Cabernets to boost their acidity (Valdiguié is known for its high acidity, and high acidity leads to long ageability in wine, hence Joe's 1970 Valdiguié's staying power) and to pump up their flavor. When Joe Swan made his Valdiguié in 1970, there were about four thousand acres of the variety planted in California, used mostly in blends, as Gallo was doing. Today, there are so few acres that the state records don't show any. Most if not all of the vines have been uprooted and replanted with hot-selling varieties like Syrah, Chardonnay, Merlot, Sangiovese, Viognier, and Cabernet Sauvignon.

I recount this story for several reasons. One, it illustrates how a wonderful wine can be personally historic for the impassioned wine lover. Two, it introduces you to my friend the late Joe Swan, an airplane pilot before he turned grape grower, who brought Pinot Noir cuttings from Domaine de la Romanée-Conti in Burgundy to this country and established them in his Sebastopol vineyard, where they became the widely planted "swan clone" (if you've had Sonoma County Pinot Noir, you've undoubtedly drunk wine from vines propagated from the pocketful of sticks Joe brought back from France in the 1960s). And three, it illustrates how fine wine is, after all, an agricultural product, grown on wine farms. Some of these farms are huge agribusiness affairs, drenched in chemicals and producing indifferent wine. But others are artisanal wine farms—small to medium-size estates, similar to the small chateaux of France—that are often farmed organically or at least in a low-input, sustainable manner. It is the latter kind of wine farm we're interested in here—especially in the surpassing quality of their wines and how they get that way.

The Origin of Wine Flavor

What causes the unique, delicious flavors of wine? Are they in the grape juice to begin with and then unlocked by the action of yeast? Or are they formed by yeast from precursors in the grape juice? Or are they something the yeast itself imparts to the wine? The answer is all of the above.

The grape berry has three parts—the skin, the pulp, and the seeds. In general, the fruit flavors, color, and aroma compounds are found in the skins. The tannins are found in the seeds—or rather on the seed surfaces. And the sugar and liquid juice are found in the pulp. These three parts of the grape are compartmentalized from one another—until crush, when the grape bunches are run through stemmer-crushers. As the name of the piece of equipment suggests, it takes the stems from the bunch and ejects them, then crushes the berries so that tannins mix with skins and juice in a big soup called must. Large wineries use fermenting vats, but for small batches, nothing beats tromping on the grapes with bare feet, which gently separates the berries from the stems and crushes them. The stems—which are also high in tannins and would make the wine too tannic if significant amounts of them were allowed in the fermenting vat—can then be removed from the crushed grape must with a pitchfork. I've done the job both with stemmer-crusher and by foot, and the very best way for the home winemaker is to open a good bottle of wine, put your spouse in the vat with the grapes, put on a recording of Sousa marches, and encourage him or her to tromp those grapes with sips of wine, shouts of joy, and lots of kisses.

The flavor-color-aroma compounds in the skins would oxidize quickly if it weren't for the tannins that are now mixed with the crushed grape juice. Tannins are used to preserve hides, and they do just as good a job preserving the fresh fruitiness of grapes. These flavor compounds are called phenolics. Some are astringent, some bitter, some sweet, some fruity, and some—anthocyanins—are the source of color as well as flavor (and even a bit of sweetness given not by sugar but by glucose-like substances associated with them). When the anthocyanins are mixed with tannins, they are called complex anthocyanins, and you can thank them for the wonderful flavors and aromas of your wine.

Esters in the grape berries are also responsible for the aromas of good wine—the black cherry, tropical fruit, blueberry, and other smells that contribute so much to the flavor of wine. There are only five tastes—sweet, sour, salty, bitter, and umami—and all the rest of what we think of as flavor in wine comes from its aromas. Esters, too, are preserved by tannins. Young tannins in newly made wine can be aggressive: mouth-puckering, astringent, and gritty. As wines age, these tannin molecules form polymers, or longer chain molecules, and give a much softer effect in the mouth. Winemakers encourage tannins to form long-chain polymers in young wines by allowing extra time—usually twenty-eight days or more—in the fermenting vat for the skins and seeds to be in contact with the new wine. This technique is called extended maceration, and wine finishes with softer, less aggressive tannins.

The yeast adds a lot to the wine during fermentation, forming volatile fatty acids that give the wine a racy, sometimes meaty aroma and flavor. When yeast cells die, they spill their cell contents into the wine (a process called yeast autolysis), and this brings toasty brioche characteristics to the wine.

In addition, terpenes are substances that are bound to the sugar in fresh grape juice. When the yeast nabs the sugar to ferment it to alcohol, the terpenes are freed and lend spicy flavors and aromas to the wine.

The yeast also changes sugar into ethyl alcohol (or ethanol), liberating carbon dioxide as it does so. If you've ever seen saccharomyces going after a vat full of sugary grape juice (ripe grapes can be 25 percent sugar or more), you'll have seen the must literally boiling with bubbles as the liberated carbon dioxide comes off. Winery workers have to be extremely careful when working around tanks of fermenting must lest they enter an area where carbon dioxide, which is heavier than air and is odorless, has pooled and driven off the air. Workers have been known to pass out from lack of oxygen and then die from asphyxiation, not because carbon dioxide is poisonous, but because it has replaced oxygen-rich air.

Other contributors to wine quality include what was formerly called *Leuconostoc oeni,* but has now had its name changed to *Oenococcus oeni.* These bacteria change malic acid—a rather harsh and bitter acid—to lactic acid, which has a softer texture and buttery flavor. This is the malolactic fermentation, and it has nothing to do with yeast; it's strictly bacterial. Winemakers add *Oenococcus* to musts near the end of the sugar-to-alcohol fermentation so the malolactic fermentation will be completed at about the same time. Otherwise the wine may decide to wait until springtime, when the weather warms up, to go through a spontaneous malolactic fermentation and push the bungs out of the barrels. Here's another area where genetic engineering is being pursued. Scientists have inserted the gene from *Oenococcus* that changes malic acid to lactic acid into *Saccharomyces cerevisiae* so that the yeast might do it all—convert sugar to alcohol and malic to lactic acid. So far, the modified yeast hasn't been able to convert enough of the malic to lactic to make the strain worth using. But that may change with time.

Alcohol itself is a good solvent and lifts the flavors and aromas to our palates and spreads them around inside our mouths. Think of a good perfume—the scent is in an alcohol base for the same reason. When wine tasters want to really get a good impression of a wine, they chew, slosh, and bubble air through the wine in their mouths so that the aromas go up the back of their noses. This is called the retronasal, and it allows for the fullest appreciation of a wine's "stuffing," as the pros call the flavors, aromas, and nuances of a fine wine.

What Vines Prefer

Now let's talk about viticulture—the farming of grapevines that give us that marvelous fruit. Most agricultural crops prefer deep, dark, crumbly rich soil churned by earthworms, plenty of water, tons of sunshine, and warm nights. But not wine grapes. The best wines are made from vines that have to struggle. Do you have poor, shallow, rocky soil that doesn't appear to be good for weeds, let alone grapes? Good. Lack of water just when the summer is at its hottest? Great. Cool, even cold, nights? Perfect! The gnarly old vines just a couple of feet tall that claw to stay alive in the dry soil on the rim of an extinct volcano on the island of Santorini in the Aegean Sea produce some of the best wines of Greece.

In Spain, Italy, and in California's wine country, it usually doesn't rain from May to October—pretty much the entire growing season, bud break to harvest. Where the climate is more forgiving, such as France, growers stress their vines by forcing them to compete intensely for whatever sustenance is available, cramming them together in tight rows—up to 11,000 vines per acre compared to about 680 in a typical old-fashioned California planting—and by planting them on the stoniest, chalkiest, driest soils they can find. In Bordeaux, the finest red wine grapes are grown on the gravelly slopes of the gently rolling hills, the lesser white wines planted in the more fertile valleys.

But why would harsh conditions favor fine wine?

The answer lies in the way vines respond to environmental stress. Grapes are tenacious, able to withstand the annual summer droughts of the Mediterranean climate and still ripen into a generous crop of fruit. They do this by plunging their roots deep into the earth,

six feet or more straight down, where enough residual soil moisture from winter rains exists to keep them functioning. Such deep roots allow them to absorb minerals from rocky subsoils that strengthen them and lend nuances of flavor to their fruit. Even more important, vines respond to lean soil and lack of water by keeping their grape berries small. Herein lies one of the main keys to wine quality.

Most of the color, aroma, and flavor compounds in grapes are found in the skins. The center, fleshy part of the ripe berry is mostly water and sugar and tannins that coat the seeds. With lush conditions, grape berries grow large and watery. Small berries have a greater ratio of skin to juice than large berries, and therefore concentrate their flavor and aroma characteristics in much smaller amounts of juice, which translates into powerful, rich, luscious wines. So it behooves fine wine growers to plant their vines in places where few other commercial crops will grow.

Fine wine grapes are tough plants—as long as the temperature in the winter is mild. Severe frost—anything much below 0°F—kills the canes and buds. If the frost penetrates into the ground deeply enough, it can kill the roots.

And so fine wine culture has developed in those parts of the world where hot, dry summers and mild winters are the norm. Because grapevines are dormant in the winter, they are only in leaf and fruit during those hot, dry months. As anyone who has lived on the East and West Coasts of North America can attest, the insect populations and densities in the two places are as different as night and day. Summer in the wet, humid East means insects and lots of them, rots, molds, mildews, and more. On the West Coast, people often don't need screens on their windows and deck living is the norm owing to the lack of insects. Which means that small artisanal wine grape growers in California, at least, have traditionally gone easy on the chemicals, even if the big wine producers have in the past made grapes one of the most heavily sprayed crops.

A decade and more ago, most vineyardists used chemical fertilizers and nasty chemical fungicides like captan, ferbam, and methyl bromide, applied either to the leaves of the vines to thwart mildew, anthracnose, and botrytis rot, or to the soil to conquer root rots. Many growers routinely used chemical pesticides as a preventive against insect pests, whether the pests were in the vineyards or not. Many used herbicides in the rows to suppress grasses or broad-leaved weeds. The idea was that only the vines were allowed to grow in the vineyard—everything else had to die. But sterilizing the environment caused more problems than it solved. The chemicals were environmentally expensive to make, taking great quantities of energy in their manufacture. And they were expensive for the growers to buy, raising costs. They persisted in the soil, creating lifeless dirt that, exposed to heavy winter rains, eroded at ever more rapid rates, clogging streams with silt and hastening the eutrophication of ponds and lakes. These chemicals entered local ecosystems, damaging the health of the flora and fauna exposed to them—including farmworkers, neighbors, and customers. The chemicals killed almost all the insects—especially beneficials, which as a class are most susceptible to pesticides. The ones they didn't kill were those few pests that mutated to withstand the chemical assaults. These insects bred new generations of resistant pests, and the growers had to resort to new, and usually more toxic, chemicals to kill them. Which led to more insect mutations and resistance in an ever-intensifying spiral of poison, death, and ecological havoc.

But some growers, the organically minded ones, didn't use chemicals in their vineyards at all, and they did just fine. And for the past fifteen years or so, fine wine grape farming in California has been turning organic at an ever-increasing pace. Most of the wineries in Napa, Sonoma, and Mendocino counties have instituted some organic techniques, even if they reserve the right to use a chemical to bail themselves out of a crisis and therefore don't qualify for organic certification.

A Strong Environmental Consciousness

I recently polled winemakers and vineyardists in these three counties, and every single one said they believed in a natural, sustainable way of farming that protects the environment. Robert Mondavi Winery's line of wines "are blended from naturally farmed grapes," according to a Mondavi spokesperson. The winery joined Earthbound Farm, a large grower of organic produce in Carmel Valley, California, in an event celebrating "two of life's greatest pleasures—naturally farmed fine food and wine." However, "natural" and "sustainable" farming have no meaning defined by law, and are certainly not synonymous with organic farming. Lately the public has been clamoring for anything organic. Organic products have enjoyed a 20 to 25 percent annual growth over the past decade, while traditional grocery sales are growing at only 2 to 3 percent, according to the Hartman Group, a research company based in the state of Washington. Some of this demand is driven by a sophisticated understanding of and belief in organic methods, but most is likely the result of the public's leeriness of genetically altered crops and pesticide-laden foods, and its genuine desire for a safe, wholesome food supply produced in a safe, wholesome environment.

The demand has resulted in a boom in organic viticulture, where total organic acreage in California has zoomed from 178 acres in 1989 to over 15,000 acres today. Big wineries are slowly moving toward organic culture, even if most are still at the stage of sustainable agriculture that minimizes but doesn't eliminate chemicals. Tim Mondavi, winemaker at Continuum and son of the late Robert Mondavi, who's been proselytizing for a more natural viticulture for close to twenty-five years, says, "The wine country is a beautiful place and it's up to us to protect it." At Gallo of Sonoma, viticulturist Jeff Lyon says, "We grow about two thousand acres of fruit in Sonoma County using IPM [integrated pest management, which uses the least environmentally disruptive pest controls first before moving on to toxic chemicals]. We also have a full set of weather stations that warn us when conditions are right for outbreaks of disease like mildew and pests like spider mites. So we can take measures only as needed, not as routine applications of chemicals. And soil is key, so we use permanent cover crops to increase the diversity of life in the vineyards and improve the soil naturally."

Cardinale Winery in Oakville, Napa Valley, recently achieved certification for twenty-two organic acres. Both Kendall-Jackson Wine Estates and Jackson Family Farms (the two entities that wine behemoth Kendall-Jackson has split into) have announced a ban on selected pesticides, including methyl bromide, Omite, simazine, and Karmex, in its vineyards worldwide.

Jean-Charles Boisset, the scion of a Burgundian wine family with extensive vineyard and winery holdings here and in France, is adamant about turning acreage under his control into an organic and biodynamic culture. He has aggressively bought up some major wineries in Napa and Sonoma counties and moved them into organic and biodynamic production. He has set up his headquarters at Raymond Vineyards in the Napa Valley, where he has created a Theater of Nature, an educational display exhibiting the great wheel of life as it affects grape growing and winemaking. Don't miss it if you visit the Napa Valley.

Smaller wineries with strict organic practices find sales booming, too. Jonathan Frey of the rigorously organic Frey Vineyards in Mendocino County says, "We've doubled production to thirty thousand cases just in the last five years." Today, more than seventy vineyards in California are certified organic. And the phenomenon not only keeps growing, but is worldwide, with more than four hundred wineries producing more than two thousand organic wines, mostly in France but also in places as far-flung as Chile. The Carmen Winery, for instance, produces 100 percent organically grown Chardonnay and Cabernet Sauvignon in the famed Maipo Valley there.

While peasant winemaking in most European countries has long been natural—that is, done the old way without factory-made chemicals—and qualifies as organic with a small "o," it usually doesn't qualify as organic with a large "O"—that is, it doesn't follow an organic farming program that fulfills the requirements for certification. France has led the way in true organic wine production through several umbrella organizations, especially the National Federation for Organic Wine, created in 1998, and the National Observatory for Organic Farming, created in 1996, to keep records on the number of organic farms, their acreage, and produce. In addition, there are three independent third-party organizations in France that certify vineyards as organic (or *biologique*, as the French term it) according to standards approved by the French Ministry of Agriculture. They are Ecocert, Biofranc, and Qualité France. The organic (and biodynamic) movements in French viticulture have been exploding of late.

The Genesis of Organic Winemaking

The organic wine industry in the United States began in California in 1956 when Nick Lolonis of Redwood Valley, Mendocino County, began to farm his family's estate vineyard organically. Founded in the 1920s by his father, Tryfon, a Greek immigrant, the original vines still exist, but the estate has grown to more than three hundred acres. Tryfon's three sons, Nick, Petros, and Ulysses, share vineyard and winery management today, and Petros's son Phillip handles the marketing of the winery's 27,000 cases.

Lolonis's commitment to organic viticulture stems from the old-country style of natural farming that Tryfon established and his sons have continued. "Uncle Ulysses is sixty-eight and has known these vines all his life," Phillip says. "He will walk through the vineyard with me and say, 'These vines need to be watered.' I ask him how he knows, and he says, 'The vines are telling me.' His intuition and life-long familiarity with the vines is amazing. And he won't let me touch a vine—he says I don't have enough gray hairs yet."

Today's Organic Viticulture

While Lolonis's viticulture started as old-fashioned farming that simply abjured agricultural chemicals, today it is truly organic, including the use of cover crops in the rows, natural pest management, and fertilization with composts and manures. A key factor in organic grape growing is the use of cover crops such as clover, vetch, and legumes for soil improvement, and umbelliferae (plants like dill and wild carrot that form umbrella-shaped flower heads), which are prime food sources for beneficial insects. Another organic technique includes allowing some areas near the vines to grow wild. These wild patches provide food and habitat for the indigenous fauna, including beneficial insects, which add a healthy diversity to a vineyard's ecosystem. "We have bats, eagles, hawks, and barn owls, all without putting up nesting boxes," Phillip Lolonis says. And the wild blackberries that grow around the farm's waterways, and are a prime breeding ground for beneficial insects, have been used "for Grandma's blackberry pies since I was a kid."

If there are any gaps in the protection afforded by natural beneficial insects like green lacewings, things are helped along by the monthly release of twenty-five gallons of ladybird beetles, better known as ladybugs, during June, July, and August—that's about five and a half million predators looking for aphids, spider mites, and other pests. Lolonis also releases praying mantises, although these indiscriminate and voracious predators will eat whatever they can grab—beneficials or pests.

One stubborn problem for organic grape growers is the ubiquitous presence of phylloxera—a form of plant louse that eats away the roots of *Vitis vinifera,* eventually killing the vines. The world's fine wine grapes are all cultivated varieties of *Vitis vinifera,* a botanical name that means "wine-producing grape," and whose ancestral home is variously placed in the mild climate zones of the Caucasus and the former Soviet Republic of Georgia or thereabouts. It's thought that during the great Indo-European migration that took place after the last Ice Age, about 10,000 years ago, tribes from the Caucasus carried their vines westward as they moved into the Mediterranean area. Evidence of winemaking in the Near East goes back eight thousand years. The Greeks took their vines to Italy about 1000 BC, and shortly thereafter, vine culture reached France and Spain. Through more than a hundred generations, people have been selecting exceptionally flavorful varieties of *Vitis vinifera,* and today's Cabernet Sauvignon, Cabernet Franc, Merlot, Gewürztraminer, Riesling, Pinot Noir, Chardonnay, Pinot Blanc, Syrah, Petite Sirah, and hundreds of other varieties are all types of this wild vine that still grows in the fields and forests of the Caucasus Mountains. Researchers have made wine from the wild, ancestral *Vitis vinifera,* and it reportedly makes a coarse but pleasant wine, one that would be infinitely pleasurable to ancient nomads who had no other wine to drink.

Because phylloxera is native to North America, however, wild North American grapes have evolved resistance to this pest. Their defense is a thick, corky bark that covers their roots, through which the root louse can't penetrate. Almost all the Vitis *vinifera* planted in the United States and Europe is now grafted to phylloxera-resistant native American grape rootstock. A question naturally arises about phylloxera and Lolonis's vineyards. How is it that just about every vineyard in the Napa Valley and many in Sonoma County had to be replanted in the 1990s because of outbreaks of the phylloxera root louse, and yet Lolonis's seventy-plus-year-old vineyards have not? One reason is that when California viticulture exploded in the late 1960s through the 1980s (actually, it's still expanding rapidly), many growers planted vines grafted to a rootstock called AxR, which was recommended by the University of California at Davis, the nation's premier school for viticulture and winemaking. Unfortunately, AxR is not resistant to phylloxera (putting a nice dollop of egg on the face of UC Davis), which caused untold woe and expense when beautiful vineyards just entering their prime years showed decline and had to be ripped out. The old Lolonis vines may have been planted on an older rootstock called St. George, which was indeed resistant to phylloxera. But another reason may be found in the organic treatment of the soil.

In a recent survey of California vineyard soils, Frey Vineyards soil was found to be the most resistant to phylloxera. Frey is located not far from Lolonis, in Redwood Valley, a small, sleepy village that's the hotbed of organic viticulture in California. Frey was founded by physicians Paul and Marguerite Frey in 1961, with their first grapes planted in 1967. In the 1970s, their son Jonathan studied with organic guru and soil specialist Alan Chadwick at the University of California at Santa Cruz, and eventually converted the family's seventy acres to organic, and then biodynamic, culture. Chadwick, a charismatic Englishman, was responsible for touching off a huge interest in organic farming and gardening in Northern California. All Frey's acres are Demeter-certified (Demeter is the international certifying body for biodynamics), and all the grapes they purchase off the farm are either Demeter-certified or certified organic by the California Certified Organic Farmers. The vines are dry-farmed (without irrigation), the soil is improved with composts from a neighbor's five-hundred-cow dairy herd, and clover, vetch, barley, rye, and mustard are used as green manure and cover crops.

In the Napa Valley, John Williams, the owner and winemaker at Frog's Leap

Vineyards, claims that organic soil improvement revived a phylloxera-plagued vineyard he bought in the early 1990s. It had been given up as dead by the previous owner, who didn't want to go to the expense of replanting (about $50,000 an acre currently). Williams has dry-farmed it organically since, using composts made from grape pomace—the skins, pulp, and seeds left over after the juice is pressed out. The vineyard came back to life and the vines were saved.

The Core Principle of Organic Farming

A healthy soil is one that is home to a great diversity of plants and animals, from the large to the microscopic, by virtue of its content of actively decaying organic matter. Organically treated soil may be resistant to phylloxera and other pests and diseases because the good microorganisms whose numbers have become myriad in the composting process have colonized all the ecological niches, leaving no room for pathogens to take hold and multiply. Does this idea sound familiar? If you remember the discussion of resistant soils, a healthy intestinal ecosystem in the human alimentary tract, and the actions of bacteria and their bacteriocins (see Chapter 2), you'll see that the same principle is at work: The more diverse the ecosystem—that is, the more hospitable the system is to a large number of creatures—the healthier it is.

The USDA's Final Rule on Organic Wine

The U.S. Department of Agriculture has promulgated the rules on what's organic and what's not, and the definitions regarding organic wine are set in law. It wasn't an easy process for organic grape growers or winemakers, especially when it came to the knotty problem of sulfites. Potassium metabisulfite is a compound routinely added to grape juice to inactivate or kill spoilage organisms and as an antioxidant to preserve wine freshness. When it's added to the liquid juice, it dissociates and forms free sulfur dioxide, a radical that will bind with and destroy spoilage organisms, primarily bacteria that cause off flavors. At high enough levels, it will do that to yeast, too, but at the low levels used in organic winemaking, usually 100 parts per million or less, it inhibits bacterial spoilage. Just as important, it has the ability to bind with oxygen, and thus functions as an antioxidant. (An oxidized wine develops off and sherry-like flavors, seems old and tired, and loses its fresh and fruity flavor.) Winemakers consider the use of sulfites, as the compound is called, essential to prevent spoilage and preserve the wine's integrity. Sulfites are a naturally occurring substance in the human body, by the way, which exists at much higher levels than those found in wine. Some people claim that wine—red wine, especially—gives them a headache, and blame sulfites. While a very few people may be sensitive to sulfites, most studies show that people are sensitive to other substances in the wine (alcohol, for instance, or tannin).

At first, the National Organic Standards Board, the group charged with defining organic for the USDA, proposed that sulfites be allowed in organic wine. The USDA changed that, disallowing their use in any wine labeled organic, and furthermore disallowing any use of the word "organic" on wine labels if sulfites were added, even if the grapes were organically grown. The wine industry in general, led by the organic segment, reacted vehemently against that rule. It believed that the USDA was trying to marginalize the whole sphere of organic wines by limiting the name to those few producers—and there are just a handful—who use organic grapes and refuse to sulfite their wines. The Organic Grapes into Wine Alliance, Bonterra Vineyards, the Organic Wine Company and Chartrand Imports (importers of organic wines from Europe), Fitzpatrick Winery, Badger Mountain Vineyard, and many other organizations fought the USDA on this and won, partially.

As the final rule now stands, only wines made from organically certified grapes and made without added sulfites can be labeled as "organic wine." Wines made from organically grown grapes with the addition of 100 ppm sulfites or less can state "made from organically grown grapes" on the label. That seems fair enough. But there may be more wrinkles. Brian Fitzpatrick of Fitzpatrick Winery in Somerset, California, told me that "the interpretation of the labeling is still not clear. For instance, there will be a 100 Percent Organic label—certified grapes and no sulfites—and there will be a USDA Organic seal on the bottle. Then there is a different USDA Organic seal that ensures that at least 95 percent of the product is made from organic ingredients." There will be a USDA seal on that bottle, too. Then, he said, there is a label where only 70 percent of the product must be from organic sources: the words "Made from Organically Grown Grapes" may be allowed on the label, but the USDA Organic seal may not be used. The problem, according to Fitzpatrick, is that wine made entirely from organic grapes with sulfites added (by far the largest amount of wine falls into this category) may be treated under the 70 percent rule and not gain the USDA seal, even though it meets the criteria for the 95 percent rule. "Does that help?" Fitzpatrick asked. Well, it sounds as if the definition of what's organic has fallen into the hands of USDA bureaucrats, with predictable results. I suggest you simply read the label closely. It will tell you the status of the grapes used to make the wine. It may even say "Vegan Wine." That's wine made with no animal-derived fining agent, such as egg whites or isinglass (the latter derived from fish). Fining agents carry a slight electrical charge on their surfaces that draws cloudy particles and impurities to them. These clump together and precipitate out of the wine, leaving it clear. Some winemakers believe that fining removes some of the nuance of the wine and don't do it. Most use fining agents. If vegan wine is fined at all, it is with bentonite clay, a natural mineral product.

Yeast and Wine

Wine was originally made by spontaneous fermentation; that is, whatever yeast and bacteria happened to colonize the grape skins or came floating along and landed in the grape juice fermented that juice into wine. But in modern times, vintners have begun using pure strains of specific types of yeast. This enables them to better predict what the finished wine will be like and how the fermentation will proceed, but it also means the loss of the complex flavors created by a diversity of microorganisms. "Now we're looking at using natural mixes of organisms again not only for their ability to add complex flavors to wine, but because a healthy mix of microorganisms reduces the chance of spoilage," says Tom Tiburzi, winemaker at Domaine Chandon in Yountville, California. "In Switzerland in the

1970s, cheesemakers started using pure strains of starter bacteria and they suddenly found the holes in the cheese were gone. People wanted the holes. So they needed to go find the wild microorganisms that made the holes in the cheese and reintroduce them into the process."

While *Saccharomyces cerevisiae* and *S. bayanus* are the primary fermentation yeasts, many others can come into the vats on the skins. *Kloeckera, Metschnikowia, Kluyveromyces, Candida,* and *Pichia* are all common genera of yeasts that may take part in fermentation, but they tend to die off as the *Saccharomyces* species gets going and starts building up alcohol levels. Alcohol, being a by-product of yeast metabolism, is toxic to yeast when it reaches certain levels. Some yeast die off when alcohol reaches just 6 or 7 percent; others can withstand a few percentage points more. *S. cerevisiae* and *S. bayanus,* however, can withstand alcohol levels up to 17 percent or so before dying off, and that's why they are used as pure strains.

Tiburzi told me a story that points up the problems with trying to manipulate yeast instead of letting nature take its course in the production of wine. Yeast produces hydrogen sulfide as a precursor to amino acids in wine. Now, hydrogen sulfide is a gas that smells like rotten eggs—not something you want in the nose of your Merlot. During rapid fermentation, the hydrogen sulfide disintegrates as the sulfur in its molecule is used to build amino acids—which are good things in wine. However, toward the end of the fermentation, when the yeast are weak from having used up most of the nitrogen in the must (must is a winemaker's term of art and refers to the fermenting crushed grape and juice), they leave much of the hydrogen sulfide unchanged instead of transferring it into amino acids, which leaves the wine smelling awful. So scientists have taken to adding yeast nutrient (nitrogen) to fermenting musts in the form of diammonium phosphate (DAP) so that the yeast will stay vital long enough to clean up the hydrogen sulfide.

So far so good. But in a related matter, yeast also excretes urea during rapid fermentation. Urea in a hot wine tank can be transformed into ethyl carbamate. Ethyl carbamate is a close relative of carbaryl—which is a pesticide known as Sevin. Because ethyl carbamate is a suspected human carcinogen, several countries have imposed restrictions on the amount permitted in wine. The saving grace is that yeast will use up the urea at the end of the fermentation, taking its molecule apart to get at its nitrogen (remember, nitrogen is a yeast nutrient). Therefore, it's no longer available to be transformed into ethyl carbamate. Except, and here's the kicker, *except* when diammonium phosphate is present. Then they'll go for the nitrogen in the DAP and the unused urea can be transformed into the carcinogen. The simple solution would be to make wine naturally and not add diammonium phosphate to it. But here come the genetic engineers again. According to Linda Bisson, a professor in the Department of Viticulture and Enology at UC Davis, they are trying to "make wine safer for human consumption. . . . An example is our efforts to genetically engineer strains [of *Saccharomyces cerevisiae*] that reduce the appearance of ethyl carbamate in the fermentation environment."

That's noble, but it reminds me of the situation with nitrogen-fixing soil bacteria, whereby these little farmer's helpers will take nitrogen from the air and turn it into fertilizer for crops—*except* when the soil is flooded with chemical nitrogen. Then they lose their ability to fix nitrogen. Thus the farmer pays good money for an expensive fertilizer that turns off a natural mechanism that was giving the same fertilizer to him for free. Similarly, there are natural, nonchemical ways to control weeds in crops. Herbicides are used, but they can damage the crops as well as the weeds. Instead of simply using natural weed controls, scientists are genetically engineering crops to withstand the chemical assault of herbicides.

Saccharomyces is added as a pure strain to grape juice the way we add it to flour and water to make bread dough—and just being in a winery will seed the grape juice with enormous numbers of these yeast cells. "Grape juice is a wonderfully selective medium for *Saccharomyces*," says David Mills, assistant professor of microbial ecology and bacterial genetics at UC Davis. "The winery has billions and billions of *Saccharomyces* cells all over the place from previous fermentations. Therefore, carrying juice through the winery without it getting inoculated with resident

Saccharomyces is problematic. I tell my students that it would be like trying to walk through a car wash without getting wet."

A yeast named *Brettanomyces intermedius* is a problem in wineries, because, although it takes part in alcoholic fermentation, it yields haziness, high concentrations of acetic acid (a vinegary taint known as volatile acidity), and a particularly unpleasant smell called "mousiness." Brett, as it is called, is definitely a spoilage organism, but cleanliness in the winery and the use of 100 ppm sulfites in the wine usually control the problem.

Winemakers today are increasingly allowing fermentations to proceed with the yeasts that occur naturally on the grape skins. They feel that each of the many yeast species adds its certain "something" to the flavor complexity of the wine. While most of these natural yeasts die off at lower alcohol levels, winemakers keep the fermentations going to dryness, if that's their goal, by adding commercial wine yeasts that blow through the remaining sugar quickly and completely as the natural yeasts die off.

Fiddling with Yeast Genes

What if the grapes are grown organically and no sulfites are used but the juice is fermented by yeast that has been genetically modified? Could it be labeled organic wine? Technically, no. But in the United States, the wine could still be labeled organic because there is no way to know whether a GMO (genetically modified organism, yeast in this case) has gotten into the fermenting must. The government does not require that products containing GMOs be labeled as such. That's not true in other countries, especially Europe and South Africa, where there's a large wine industry and a growing distaste for anything with GMOs in it.

In Australia and the United States, much research is being done on modifying wine yeast (our old friend *Saccharomyces cerevisiae*) genetically, and consumer fears about the potential dangers are generally dismissed by researchers. Professor Linda Bisson recently wrote, "Consumers raise another issue, that the process of genetic exchange and mutation is unnatural and therefore undesired. This view is of course not correct . . . The risks to human health and well-being of the genera-tion of such modified strains of *Saccharomyces* for food and beverage production are minimal."

Is it too far-fetched to suppose that a genetically modified strain of *Saccharomyces* could be engineered against which the human immune system has no defense? Professor Bisson's blanket statement, "This view is of course not correct," sounds like the kind of hubris that the fates love to humble.

Scientists know that yeast fermentation of grape juice creates esters that give those wonderful fruity aromas and flavors to wine, such as apple, banana, fruity-flowery, and tropical fruit aromas. The yeast performs these miracles by manufac-turing enzymes that work on grape juice to produce acetate esters. In one study, scientists investigated the feasibility of improving the aromas of Chenin Blanc by overexpressing the gene within the genetic structure of *S. cerevisiae* that makes an enzyme that creates a fruity ester. Usually this enzyme's manufacture by the yeast cell is controlled naturally by an on-off switch in one of the yeast's genes, with the switch setting determined by environmental conditions. If conditions are right, the yeast makes the enzyme; if conditions aren't right, nature in her wisdom shuts off the gene. Along come the genetic engineers, who say, "Let's throw the switch on permanently and see what happens." Sort of like the Sorcerer's Apprentice and the brooms and buckets. They tried it—and what happened? Here is the study's conclusion: "Overexpression of acetyltransferase genes may affect the flavor pro-files of wines deficient in aroma, thereby allowing for the production of products maintaining a fruitier character for longer periods after bottling." In other words, it works.

In another example from the study: "Researchers at the Australian Wine Re-search Institute are using genetically engineered yeasts to increase the flavor of wines. Australian viticulturists like their grapes to fully ripen so they impart more flavor to the wine, but ripe grapes contain more sugar, which, when fermented, leads to wines containing higher alcohol concentrations. High alcohol wines tend to exhibit a 'hot' flavor, which winemakers, and more important, consumers, don't like. AWRI has genetically modified yeast to produce more glycerol and less sucrose. This will reduce the alcohol levels." In other words, this GMO turns some of the carbon, hydrogen, and oxygen atoms in the grape juice to glycerol rather than to grape sugar. Yeast turns sugar to alcohol, but not glycerol, hence final alcohol levels are suppressed.

In a final example, out of many, many more: "Wine strains of *Saccharomyces* producing bacteriocins have also been constructed," according to Professor Bisson. "Bacteriocin production by yeast strains offers many advantages over the current practice of use of sulfur dioxide." In other words, genetically engineer a strain of killer yeast that will knock out all diversity in the must, save for itself.

So what's wrong with tinkering with the genetic control panels in yeasts to make them do what we want them to do?

Plenty.

It must be realized that naturally occurring genes represent a series of choices made over millions of years through natural selection, with nature's overriding concern for the preservation of life. Each organism, large or small, is a whole system, with an integrity made up of a maze of complicated interactions with its environment. It's genetically structured for optimization of its survival. It has an ecological niche that represents its place and its job in the interwoven web of life. It's a cog in the gears of life on earth. With genetic engineering, human beings have found their way into the control room of life, and we are—as you would expect from primates like us—monkeying around. We're pulling levers and pushing buttons, just to see what will happen. Sooner or later, we will modify a gene that will become a rogue in the environment (it has already happened with the release of Bt corn pollen into the farmlands of the world). If someone asked me what will life be like a hundred years from now, I'd have to say that humanity will be spending an enormous amount of resources tracking down and eliminating rogue genes, with catastrophic impacts on the environment and the natural ecology as we do so.

"The genie is already out of the bottle," said Neil E. Harl, a professor of agriculture and economics at Iowa State University, speaking of GMOs. "If the policy tomorrow was that we were going to eradicate GMOs, this would be a very long process. It would take years, if not decades, to do that."

A Visit to a Small Winery

In 1972, I made my first batch of wine, from grapes being thrown out of a local grocery store because they were too old. "Oh," I said to myself in my ignorance, "I'll make wine from them," and took them home. I tried to make wine using bread yeast, which didn't work very well. The wine was undrinkable. I poured it out. Then I tried buying grapes from California, shipped to the New York area by rail. They were crushed into forty-pound boxes and were already turning to vinegar. Then I realized that it takes fresh grapes to make good wine. So I planted a small vineyard, mostly Chancellor, a French-American hybrid that could withstand Pennsylvania's tough winters. The wine wasn't bad, but it wasn't world class. I decided to move to Sonoma County in the Northern California wine country.

A month after my transcontinental move, I ran into a group of guys who made garage wine. Tasting it, I saw right away that they made it right, from great grapes, using French oak barrels. One of the fellows was moving away, and so a spot in the four-member group opened up for me.

Soon I found myself high on Sonoma Mountain in Dave Steiner's Cabernet Sauvignon vineyard, helping to select and pick grapes. Some of the grapes had a vegetal flavor, like asparagus, that signaled they were not ripe and probably would never become fine wine. Other grapes in a sunnier location tasted wonderful—clean, fresh, bursting with fruit flavors. So we convinced Steiner to sell us a ton of the clean-tasting fruit that we selected from various parts of the vineyard. Because he knew our wine (the group had been making it from his grapes for years before I showed up) and thought it was as good as if not better than any of the commercial wineries that bought his grapes, he agreed. And all we wanted was a ton—the production of about two hundred vines out of the many thousands in his twelve acres of vines. We filled two half-ton fermenters—open-topped plastic tubs four by four feet square, about three feet deep, slipped into wooden frames to support them and loaded onto the bed of a pickup truck amid a cloud of yellow jackets, which follow the grape gondolas in wine country like camp followers after Caesar's legions. "A few yellow jackets in the wine are a quality point," we'd say.

Back at the garage we used as a winery, we picked through the grape clusters to reject any rotten or green berries, and tossed the ripe bunches into a stemmer-crusher. This small electric motor-driven machine sat on two boards laid across the top of an empty half-ton fermenter. Greenish-brown grape stems came twisting out of the end, while lightly crushed grapes fell into the vats below. When the grapes were all crushed, we covered them with a piece of plastic sheeting to keep out the fruit flies (fruit flies carry *Acetobacter* bacteria that make vinegar, not wine) and went home for a day. Allowing the must to sit uninoculated for a day is called a cold soak and is very beneficial to the finished product. The next day, we stirred two ounces of dried wine yeast (originally we used Montrachet yeast, but later changed to the gentler-acting Prix de Mousse) into a jar of the grape juice in the fermenters and gave the yeast time to dissolve and wake up. Then we poured half the jar into each fermenter. Before going home, we punched down the cap. Punching down has to be done at least twice a day. Grape skins float, and if left alone, they'll become a breeding ground for spoilage organisms. Punching down submerges the cap of grape skins under the surface of the must, wetting them and keeping them moist. For our punching-down tool, we used a flat metal stainless-steel plate we'd welded to the end of a stainless-steel pipe.

Twice a day (or more often if we were nearby) we'd take turns visiting the garage, removing the plastic, punching down the cap, and replacing the plastic. Within three days, the vats came to a rapid boil. The sweet, yeasty aroma of the fermenting must greeted us as we walked up the driveway to the garage. Fruit flies began to multiply, but only a few could reach the must, and a few aren't really a problem. "A few fruit flies are a quality point," we'd say.

After about four days, we'd inoculate the must with *Oenococcus oeni*, the malolactic bacteria. The malolactic fermentation softens the harsh acidity of the wine, and if not done during the primary yeast fermentation, can happen spontaneously later on, even when the wine is in the bottle. When that happens, the corks get pushed out and the wine spoils, so we always made sure the must went through malo at the end of the primary fermentation. After five or six days, the must would cease its vigorous seething and quit working. A hydrometer showed us whether the fermentation was finished—when all the sugar is changed to alcohol, it gives a negative reading. Even though the fermentation was over, we still had to punch down the cap. But because clouds of carbon dioxide were no longer boiling off the surface of the must, preventing the fruit flies from entering, we knew the vats needed added protection. So we bought two tanks of commercial carbon dioxide and ran clear plastic hose from the tank valves to the plastic sheets, punched little holes in the sheets, and inserted the hose into the holes. Then we'd open the valve just a crack so a barely audible amount of gas could escape and reach the vats, keeping a protective blanket of carbon dioxide over the must. At this stage, too much air reaching the must can cause the alcohol to transform once again— this time to vinegar. By keeping the must away from air, the process stops at alcohol—at wine, that is, for that is what the liquid in the vats had now become. We'd try a sip, but it was usually chalky with yeast and not very appealing at this young stage.

After about three weeks, the cap of skins had yielded up all they could give to the wine, becoming paper thin and sinking through the liquid. That was our signal to press the new wine off the skins. We had a big hand-operated basket press that we'd line with a porous mesh plastic bag. We scooped up buckets full of must from the fermenting vats and poured them into the bag. Free-run juice would pour out onto the press's platelike platform, then through a spout into clean buckets. When a bucket filled with wine, we'd carry it to a barrel and pour it in using a big, wide funnel. When the mesh bag was full of sopping wet skins, we'd slip the ratchet-operated follower onto the center spline, insert the long metal handle, and work it back and forth. Each back-and-forth motion produced a satisfying click-clack, and screwed the follower another click down onto the bag full of juicy skins. It took about five minutes to press the bag firmly enough to get out most of the wine. Then we'd remove the follower, empty the bag, reinsert the bag in the basket, and start pouring in more buckets of new wine.

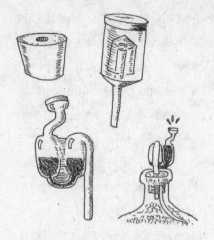

Clockwise from top left: Bung with center hole to accept an airlock. Two-piece plastic airlock. Airlock set-up on a five-gallon carboy. One-piece plastic airlock showing proper fill of water.

In this way, we soon filled two sixty-gallon French oak barrels with our young wine. The bungs on top of the barrels can be fitted with airlocks if the wine is still fizzing (airlocks allow gas out but no air back in) or with silicone stoppers that fit snugly in the bung holes. Since we did an extended maceration of about twenty-eight days, the wine was completely dry (all the sugar fermented) and so we generally used the silicone bungs. After cleaning up, our work was done for the day. It was usually about late October to mid-November when we pressed the wine off the skins.

By late November or the first week in December, we'd rack the wine off the gross lees. This means that we'd pump the wine from the barrels into large plastic containers until we saw the chalky lumpy residue that had accumulated in the bottom of the barrels start to enter the clear plastic hoses attached to our pump. At that point, we'd carry the barrels outside and turn them bung hole down so all the gunk would run out, then wash them out thoroughly with a hose and return them to their places in the garage. Then we'd pump the wine from the containers back into the barrels. The wine gets a good aeration during a racking, which helps clarify it and is quite beneficial. But once back in the barrel, the wine is kept away from any contact with air, which encourages spoilage organisms to grow. The Cabernet starts tasting pretty good at this stage, and we always made sure to taste a quantity of it as we worked.

Two more rackings, around the first of the year and again in March, removed any more sediment. We never filtered our wine (never needed to) and never fined it. Fining is a process that clarifies wine that may be a mite cloudy. But it also strips some flavor and color components from the wine. Our wine always became perfectly clear by the third racking.

After that, the wine remained in the barrel for another year. The only job was to give it a small dose of sulfites to protect it, and occasionally take out the silicone bung and top it up with wine so there was no air space inside. When the wine was almost two years old, usually in August or early September, it was time to bottle, which would not only get the wine off the oak (as time in barrel is called), but would free up the barrels for the wine that would be made that fall. Each year we'd discard the oldest barrel and replace it with one new French oak barrel. That gave us a new barrel and a year-old barrel to age each vintage of our wine. All new oak might make the wine too oaky. All old barrels might not imbue the wine with enough of the subtle flavors and aromas that oak gives. Having one new and one old barrel was perfect, but we made sure that all the wine was blended together before bottling so that the wine that had spent the year in new oak would be mixed with the old oak wine, balancing the flavors.

Bottling was fun—and hard work. Wine is poured into the top of a metal bottler that has five slender spigots. Bottles are placed so the spigot runs down into the neck of the bottle. The machine automatically stops filling when the wine reaches the proper level in the neck of the bottle. The bottles are then handed to the two guys running the corker. One guy—usually me—kept buckets of water full of corks so they'd be wet when inserted into the bottles. The other guy worked an Italian corking machine, a hand-operated gismo. I'd grab a bottle and put it onto the corker's stand. Then I'd drop a cork into the top of the corker. The guy manning the corker would pull down a long lever and the cork would be squeezed by sliding metal chocks and then driven into the neck of the bottle. Then—to bottle 120 gallons of wine—the process would be repeated 599 more times. It took two or three hours to cork all the bottles. Then they'd be put into cases, turned on their sides so the corks would be in contact with the wine, and set aside. Of course, we'd be tasting our new wine as we worked, listening to the local rock station, taking a break and eating sandwiches the women would prepare (when we weren't pressing them into service on the bottling line), and generally making a social event out of it.

Soon the new vintage would be under way, with the work that entailed. So it was not until the following April that we'd get to labeling and foiling the bottles. Our labels were handsome—black with silver embossing. These were applied using

WINE AND HEALTH

The current federal dietary guidelines recommend no more than one five-ounce glass of wine daily for women and two glasses for men. Since a 750-ml bottle of wine contains about 25 ounces, that would be five five-ounce glasses. If one has a glass of wine or two with lunch or dinner on a daily basis, one is a moderate drinker, and the person to whom all the health benefits of moderate wine consumption accrue.

Studies show that moderate drinkers do enjoy more health benefits than either teetotalers or binge drinkers. In this regard, wine is very much like cheese, which has a high fat content. A moderate amount of either is fine. Too much of either can lead to serious health problems. But then, no wine or cheese at all is simply too depressing to think about. After all, as the French saying goes, "Nothing equals the joy of the drinker, except the joy of the wine in being drunk." Even the Bible, 1 Timothy 5:23, tells us, "You should give up drinking only water and have a little wine for the sake of your digestion and the frequent bouts of illness that you have." My personal opinion, gained from much experience with a glass of good wine and a plate of good food, is that the drink relaxes me after a day's hard work, washes away care, soothes the work-jangled mind, and prompts mirth and good feeling. That, I believe, is the true source of wine's beneficial effects on the human organism. It's an opinion shared by many in the Society of Medical Friends of Wine, a physicians' group in San Francisco that researches and discusses the health benefits of moderate wine consumption.

a labeling bench. The cases of wine were brought forth. Each unlabeled bottle—called a shiner in the argot of the trade—was laid horizontally between two blocks of wood that held it steady. A label was zipped through a machine with a roller that revolved into a vat of heated glue, so that glue was applied by the roller to the back of the label. They were then fixed to the bottle by hand, making sure they were all on straight and in the right place. Then they were handed to the foiler. He used a machine that looked like a vacuum cleaner tank with a hole in one end. A silver foil capsule was placed over the neck of the bottle and the neck was inserted into the foiler, which had revolving rubber rings that squeezed the capsule tightly onto the top of the neck. Again, it took good quantities of our own wine, which by this time was tasting very, very good, to encourage us to finish the work.

The bottles were then placed back in the cases and a label fixed to the end of each case so we could tell what was inside. At the end, we divvied up the cases. A hundred and twenty gallons—minus a gallon we drank during the bottling—comes to about six hundred bottles. At twelve bottles to a case, that's about fifty cases— sometimes a little more, depending on how much juice is in the grapes, sometimes a little less. Ideally, each of the four guys gets twelve cases and the leftovers go into our little home winery's "library," to drink a bottle at a time, as the years go by, at our annual Talla Mena dinner. Talla Mena (named after a Finnish drinking game) is the name of our winery. Each year we take one bottle from each vintage from the library and sample them at our dinners. I think we've finally run out of the 1976— our first vintage, and a Cabernet Sauvignon that still tastes as good as the day it was made. No, wait a second. It tastes much better than the day it was made. That's the thing about wine.

PART 3

The Recipes

....

Wer nicht liebt Wein, Weib und Gesang,

Der bleibt ein Narr sein Leben lang.

(Who does not love wine, women, and song,

Remains a fool his whole life long.)

—MARTIN LUTHER

CHAPTER 11

The Dairy Ferments

Making Milk Kefir

Kefir (properly pronounced *keh-FEER* rather than *KEE-fir*) is a tangy, milk- or water-based drink fermented by a symbiotic combination of bacteria and yeast clumped together in a matrix of proteins, fats, and sugar. It's a wonderfully rich source of healthy, diverse microbes and will do you a world of good. Kefir originated in the North Caucasus region, but no one knows precisely where or when. It comes to us from the mists of time, most likely handed down through many hundreds of generations.

You can buy commercial kefir at the store, but you'll make a better version at home. My local market sells raw, whole, organic milk from pastured cows, and believe me, that makes kefir that's far better than the commercial product. However, regular milk—even fat-free milk—will make kefir. If you buy organic milk, you can be sure that it does not contain the residual antibiotics that are routinely used on cows in conventional dairies. Kefir microbes do not thrive in milk that contains antibiotics.

The symbiotic combination (or culture) of bacteria and yeast (the often used acronym is SCOBY) forms "grains" that resemble small cauliflower florets. Some scientific sources have found up to thirty different kinds of bacteria in the grains. The microbes in these grains proliferate in milk and will make a fresh batch of kefir—which is the fermented milk with the grains strained out—every twenty-four hours.

Kefir grains contain a water-soluble polysaccharide known as kefiran, which imparts a thick texture and smooth feeling in the mouth to the fermented milk. Kefiran ranges in color from white to yellow. The grains will grow over time, and can get to be the size of walnuts (although rice-size grains and sizes in between are most common). It's this gorgeous texture that partly explains kefir's exploding popularity. A store called Treat Petite in Manhattan was an early source for soft-serve frozen kefir, which isn't as sweet as yogurt. Frozen kefir is now found online and in some retail stores.

Milk kefir has a tangy flavor and a silky texture. It's rich, and it feels good just drinking it. You can add a splash of fruit juice to the kefir if you find the taste too cheesy, but many people prefer it plain because its taste fairly shouts the word "wholesome!" Its tangy flavor is versatile. It's perfect with morning cereal instead of milk, and you can use it in salad dressings, to make ice cream, to thicken and improve soup, instead of milk in smoothies, and in baking pastries and breads.

Kefir's microorganisms colonize the intestines and benefit health by protecting the intestine against disease-causing bacteria and by strengthening the diverse ecosystem of the gut, which also supports positive health. The kefiran in kefir has been shown in one study to suppress an increase in blood pressure and reduce blood cholesterol levels in rats. Kefir also contains compounds that show antimutagenic and antioxidant properties in laboratory tests, although it is not yet clear whether these results occur when kefir is drunk.

One noticeable way kefir will improve your health is to increase your regularity, lessen the need to strain at stools, and decrease any digestive problems you may have. Kefir reduces flatulence and is a wonder food for your intestinal flora. In fact, kefir microbes colonize your gut, especially the colon, and become part of your intestinal flora—part of you.

I bought a kit to make kefir from a seller on eBay for $26. The kit included a plastic bag of milk kefir grains and one of water kefir grains, a plastic strainer, and instructions for use. It has turned out to be one of the best $26 purchases I've ever made. All these tools and the grains, along with just about anything else needed to ferment or culture many kinds of food, can also be found for sale online at www .culturesforhealth.com. The eBay seller's literature said that kefir grains should not come in contact with metal because of their acidity. The acidity can react with certain metals such as aluminum, and cause an electrical gradient on metal surfaces that can harm the microbes. Glass, plastic, or ceramic is okay. So here's the equipment I use to make my daily kefir:

- A 1-quart wide-mouth canning jar
- A wide-mouth plastic funnel
- A plastic mesh strainer
- A ceramic bowl that is deep enough so the strainer doesn't touch the bottom when it's suspended on the bowl
- A plastic spoon

When I got my grains from the eBay supplier, both milk and water kefir grains were securely sealed in zip-top plastic bags. I put the water kefir grains in the fridge to prepare later (the recipe for water kefir starts on page 254). Following his instructions, I placed the milk kefir grains as is (unwashed) in the quart canning jar, which I then filled halfway with a pint of raw whole milk. I decided to use raw whole milk for several reasons. The raw milk would have all of its natural enzymes intact, increasing its ability to convert lactose to lactic acid. I guessed that the microbes in the SCOBY would be happier in raw milk than pasteurized. And I went with whole milk simply because of its completeness as the product that came from the cow as nature intended. Then I remembered that J. I. Rodale—who introduced organic gardening and farming to the nation in 1942, when he created *Organic*

WIDEMOUTH

Cover the top with a piece of paper towel and screw it down with the metal band that comes with the canning lid.

Farming and Gardening magazine—used to say that cow's milk is for calves, not human beings. But he was talking about milk, not kefir.

I covered the jar top with a piece of paper towel and screwed it down with the metal band that comes with the canning lid, although I didn't use the lid. This lets the kefir breathe and keeps out insects and dust and any odd bacteria, fungal spores, or yeasts that are floating in the air. Then the jar went into a kitchen cupboard for a day.

My first batch of homemade kefir was a revelation. The next morning I set the strainer on the bowl, took the band and paper toweling off the jar, and poured the contents into the strainer. There was some liquid whey that ran straightaway through the strainer into the bowl. I saw that if I wanted just whey for any reason, I could pour it off, but I had no use for it so I left it in the bowl. But the milky white part was a lumpy and clumpy mix of kefir grains and creamy, thick kefir. Using the plastic spoon, I worked the edge of the spoon back and forth in this milky mass, scraping gently against the plastic mesh of the strainer, and soon all the kefir had run through the strainer into the bowl with the whey. I put the strainer and the spoon into the kitchen sink. I wiped the inside rim of the quart jar with a piece of paper towel, just to keep things clean, then sat the plastic wide-mouth funnel on the jar. Figuring it would be best to get the grains back into milk as soon as possible, I poured the grains into the funnel so they plopped into the jar, and added a pint of the raw whole milk, replaced the paper towel and band back on the jar, and screwed it down. Then the jar went back into the cupboard. Now I rinsed off and dried the plastic spoon—again being a neat freak—and used it to stir the kefir and whey together to make a homogenous drink. I poured a half pint into each of two glasses—one for my wife, Susanna, and one for me. I drank my first batch straight, no fruit juice, and, surprisingly for a fussy eater like me, I liked it. It grew on me (literally and figuratively) as I drank it religiously every day, and now I absolutely love it. Making daily kefir is as easy as that. It takes just a few minutes, and there's always kefir in my cupboard.

As you make your kefir every day, you'll find that the grains grow in number, doubling in amount over a few weeks or so. As the grains increase in amount, their action becomes more vigorous and it takes less time for kefir to form. Try to keep the amount of SCOBY in your jar to about the size of a medium hen's egg or even a pullet egg. Freezing doesn't harm the microbes, so any extra you remove from time to time can be frozen in plastic baggies, given to friends, or shared with members of your local Fermenters Club (www.fermentersclub.com). Or, you can simply use the enlarged mass of grains to make larger batches of kefir to convert friends and family into kefir enthusiasts.

The vigor of the action of the SCOBY also depends on the temperature. The higher the temperature, the more vigorous the action. In hot weather or a room that's heated to 80°F or more, 12 to 14 hours will probably be enough time for kefir to form. It isn't dangerous if the ferment goes longer, but it becomes more and more sour and more whey separates out over 24 hours at warmer temperatures, resulting in a biting, thin kefir. If you have variable heat in your house (no air-conditioning), you can decide to put the morning-made kefir jar in the fridge in the evening, which will slow the action of the grains way down overnight—and also give you pleasantly chilled kefir the next morning. Or, if your room is chilly, down to 60°F or so, for sure go the full 24 hours or even longer. I find that 24 hours is perfect when the temperature is between 70 and 75°F. What you're shooting for is kefir that is creamy, smooth, and tangy, not acidic and thin.

About every three or four days I'll wash out the canning jar after transferring the day's kefir and its grain to the strainer. Kefir can be sticky, so a good scrubbing with a brush usually helps. Then I rinse and dry it so I don't return the grains to a jar with residual chlorinated tap water.

You may hear that it's necessary to wash the grains after every time you strain off the kefir. It's not. First of all, the chlorine is put into city water to kill microorganisms—not what we want at all. The water may also contain fluoride, another poisonous substance. Second, washing can damage the happy little communities of bacteria and yeasts that grow so well in the milk we provide them. In fact, not washing them allows them to reach a sort of climax state of goodness.

One day Susanna tasted our kefir and said to me, "This tastes more and more like the stuff the Russians were drinking at the Black Sea." She'd visited there for her midwifery work in the 1980s, and Russians are well known for their love of kefir. What that says to me is that our home-grown kefir grains are now reaching the kind of maturity that the Russians enjoy.

STORING YOUR KEFIR AND GRAINS

It may happen that you have more homemade kefir than you can drink right away. Just pour it into a container or jar with a tight lid and store it in the fridge for up to two weeks.

But what about the grains? What to do with them if you make fresh kefir every day, and then you have to leave town for a week or two? There is no problem. Put the grains in a jar and cover them with fresh milk. Put the jar in the fridge for up to two weeks. If you want to store them in the fridge for longer than two weeks, strain out the grains, pour out the old milk, and pour in fresh milk. They'll go an additional two weeks.

Even easier, and if you're going to be away for an extended period, they freeze well and last indefinitely. If I have extra grains or I'm going away for a while, I put the grains and some of their curds into a zip-top freezer bag, zip the top closed, mark it "kefir grains" with a permanent marker, and store it in the freezer. When I return, I simply put the frozen grains back in my fermenting jar, cover with milk, screw on the paper towel, and put it in the cupboard. They will have made kefir by the next morning.

And that way if friends ask me for some grains, I have a frozen bag ready to hand over.

KEFIR CHEESE

Little Miss Muffet had the right idea, eating those beneficial curds and protein-aceous whey. Kefir curds, as with any curdled milk, can be used to make a probiotic kefir cheese.

When you make kefir at home using kefir grains, you have three products in your fermenting jar: clear, slightly greenish whey; kefir grains that are a symbiotic combination of bacteria and yeast; and curds formed from milk solids by the grains. When I make my morning kefir, the grains are separated from the curds and whey. I mix the whey into the curds to make a smooth, silky beverage.

To make kefir cheese, you need to get the curds and whey into a bowl and the grains back into their canning jar. Do this by setting the plastic strainer on a large bowl and empty the canning jar into the strainer. The whey and some of the thick, creamy curds will go through into the bowl, leaving curds and grains behind. Scrape the grains with a spoon back and forth in the strainer, getting as much of the curds as you can into the bowl. Place the grains back into the canning jar, add 2 cups of milk, cover it with the paper towel and band, and put it back in the cupboard to become tomorrow's kefir for drinking. Using the plastic spoon, mix the curds and whey in the bowl into a homogeneous mixture.

Now cover the bowl with a clean cloth and let it sit in a warm, dry place for 24 hours. After 24 hours, the kefir in the bowl will have made a lumpy mass of curds and whey.

Line a colander with two layers of cheesecloth and set the colander over a large bowl. Pour the curds and whey into the cheesecloth, bring up the edges, and tie off the neck so that you have a bag full of kefir curds and whey. Hang the bag from the faucet in the kitchen sink or set the colander and bag far enough above the bowl so the bag doesn't touch the whey that drains off, as you want to separate the curds from the whey. You can make kefir ricotta from that whey (see pages 184–85).

When the drainage has slowed, gently squeeze the bag. Give it a gentle squeeze every half hour until all drainage stops. Turn the curds out of the cheesecloth into a clean food container, thoroughly mix in sea salt to taste or go no-salt if you prefer, and store it in the fridge. Use it like cottage cheese in salads, spread it on toast, or add dollops to cornbread batter. It tastes great.

Coconut Milk Kefir

What's the difference between coconut water, coconut milk, and coconut cream? Coconut water is the mostly clear juice that is found in the hollow middle of the coconut, poured off without processing. Coconut milk is made from shredded coconut soaked in an equal amount of water, poured through cheesecloth, and then the cheesecloth is wrung out to capture the coconut milk. If this milk is allowed to sit in the refrigerator, coconut cream will rise to the top. To make coconut milk kefir, you can either buy ready-made coconut milk at the store or make it yourself by cracking out fresh coconut meat and grating it by hand or by shredding it in a food processor.

If you have enough milk kefir grains, reserve ½ cup and make milk kefir using the rest, as usual. The ½ cup will be used to make coconut milk kefir. Be aware that it may take a couple of tries for the grains to become accommodated to the coconut milk. Don't give up. Just make another batch the next day. Soon your grains will be turning out rich coconut milk kefir that you can use as you would crème fraîche, yogurt, sour cream, or buttermilk.

Makes about 2 cups

½ cup milk kefir grains, unwashed
2 cups coconut milk

1. Place the kefir grains in a wide-mouth quart canning jar and pour in the coconut milk.

2. Lay a square of paper towel across the top and screw down the lid band (but not the lid).

3. Allow the coconut milk to culture at room temperature for 12 to 15 hours, less time for warmer temperatures, more time for cooler temperatures, until the milk is cultured and thick. If the first batch fails to culture, pour off the coconut milk and reserve in the fridge for other uses, and make a second or third batch, if needed.

4. Use a plastic strainer and spoon to retrieve the cultured coconut milk curds as you would for milk kefir, adding the grains back to the quart jar to make tomorrow's batch. If you have enough coconut milk kefir in your fridge, revert the grains to cow's milk, or freeze them in a zip-top freezer bag. To refresh the grains, let them culture cow's milk for 24 hours at least once every 3 or 4 weeks.

➤ NOTE: This last direction only applies to kefir grains that have been used to make coconut milk kefir. Regular cow's milk kefir grains are refreshed daily by their 24-hour soak in milk. But for the coconut milk grains to remain as a strong SCOBY, they need to be revivified in cow's milk every few weeks; otherwise the combination will weaken and the grains will eventually disintegrate.

Cultured Butter

Cultured butter, also called European-style butter, is simply butter made from fermented rather than fresh cream. The fermenting agents are our familiar friends, the bacilli that transmute lactose into lactic acid—precisely, Lactobacillus delbrueckii subsp. bulgaricus, Streptococcus salivarius subsp. bulgaricus, Lactobacillus acidophilus, and strains of bifidobacteria. You'll find this mix of bacteria in a typical yogurt culture, available online or simply at the market in yogurt, crème fraîche, or buttermilk. Check the labels of these products carefully to make sure there are no gums, stabilizers, preservatives, or other processing chemicals. Choosing organic versions will avoid those problems. The increase in lactic acid gives the butter, and the whey and milk solids that are left, a pleasant tanginess that fine chefs treasure. The butter also lasts longer because of its lower pH.

Just as great wine comes from great grapes, great cultured butter comes from the best cream you can find, with the highest butterfat content. If you're buying packaged cream, the butterfat content should be listed somewhere on the label. If you're buying from a farm, determine which breed(s) of cow are in the herd. Look for a dairy that is milking Jerseys for the highest fat content. Here's a rundown:

Percentage by weight of butterfat in the milk of the most common breeds of dairy cows

Ayrshire	3.9
Brown Swiss	4.0

Guernsey	4.5
Holstein	2.5 to 3.6
Jersey	4.9
Milking Shorthorn	3.8

The ideal heavy cream for making cultured butter would be raw, organic, and from pasture-fed Jersey cows. Pasteurized cream will work, but it should be organic, as conventional dairies often use antibiotics routinely, which can affect the action of the bacteria on the cream.

I like sweet (unsalted) butter, but you can add sea salt to taste. Note that this recipe is a two-day process. You'll need a food thermometer to check temperatures.

Makes about 1 pound of butter

6 cups heavy cream
⅓ cup yogurt, crème fraîche, or buttermilk
Sea salt to taste (optional)

1. Place the cream in a clean glass or ceramic container or bowl. Add the ⅓ cup of whichever starter you choose and stir to incorporate. Cover the bowl with a clean dish towel and place it where the temperature will be steady and between 70 and 80°F.

2. Check the cream after 18 hours. It should be noticeably thicker and taste slightly tangy. If not, allow it to stand for about another 4 hours, when it should have thickened. If you don't want to make butter right then, place the bowl in the fridge for up to 24 hours.

3. The cream needs to be about 60°F to separate properly. Set the cream bowl in a bath of ice water that reaches halfway up the outside of the cream bowl. Check the temperature until it hits 60°F, then remove the cream bowl but reserve the bowl with water and ice. You'll need it later.

4. You'll now need to beat the thickened cream. Use a handheld electric mixer, a mixer on a stand, or even a hand whisk. Beat the cream like you are making whipped cream. After the cream forms stiff peaks, keep beating and it will soon turn grainy in appearance, and then "break," as butter makers call the moment when the butterfat precipitates out of the buttermilk. You'll recognize it by the globules of yellowish butter fat that form. Reduce the speed of your mixer so you

Roll your butter in wax paper and twist the ends shut.

aren't throwing buttermilk all over the place, then stop beating and pour the buttermilk off the butter, reserving it. It's real buttermilk—great for pancakes and biscuits, so store it covered in the fridge after you've finished making the butter.

5. The next step is to rinse all the buttermilk out of the butter. If any is left in, the butter will spoil prematurely. Pour some of the ice water (important that it's ice water, as you don't want the butter to melt) over the butter and mash and stir the butter with a fork. Pour off the whitish water and repeat the rinsing as many times as necessary until the water is perfectly clear. Then press the butter to remove as much water as possible. If you want salted butter, add the salt now and work it into the butter thoroughly. You can now pack your butter into small ramekins, roll it in wax paper and twist the ends shut, or put it in any container of your choice as long as the butter is covered, as it absorbs refrigerator odors. If you have more butter than you can use right away, excess butter will store perfectly and almost indefinitely in the freezer.

Crème Fraîche

In the organic orchard, if you want to grow Braeburn apples, you have to start with Braeburn apples. That's because all Braeburn trees in the world—or all Honeycrisp, or Golden Delicious, or any other named variety—are pieces of the same tree that was the original of that variety. Apples don't come true from seed—so if you plant the seeds from a Braeburn apple, you don't get Braeburn trees, you get a pull of the genetic slot machine lever and an unpredictable genetic mix as a result.

It's the same when you're making crème fraîche. You have to start with crème fraîche. So buy the best and freshest that you can find. Here in Northern California, that would be Cowgirl Creamery or Bellwether Farms. There are excellent purveyors of crème fraîche throughout America. Use up all the crème fraîche except for two tablespoons, which will become the starter for your own batch. It's easy. When you get down to two tablespoons left of your own batch, start another batch.

Use as you would sour cream: with blintzes, on scrambled eggs, in pancake batter, in frittatas, on tacos, with potatoes, on fruit, and in stews and dishes like beef stroganoff.

Makes 1 pint

2 tablespoons crème fraîche
1 pint heavy cream, at room temperature

1. Place the crème fraîche in a quart-size canning jar, then pour in the heavy cream.

2. Cover the jar with a piece of paper towel and screw on the band that holds the jar's metal lid, but don't use the lid. This is to keep insects, dust, and foreign microbes out of the jar.

3. Let the jar sit for 48 to 72 hours, until it thickens (the speed of thickening depends on temperature). Remove the band and paper towel, put on the lid, screw it down with the band, and refrigerate.

Yogurt

Supermarkets devote whole aisles of cold cases to yogurt these days, but it wasn't always thus. It was really only in the 1960s that yogurt began to be popularized, and then it was sold mostly in health food stores. Today, of course, it's not only in every market in the nation, found as plain yogurt and every flavor and fruit imaginable, but we have those chains of self-serve frozen yogurt dispensaries, with the product usually flavored artificially and containing who-knows-what-all.

Back in the day, forty years ago, a lot of folks made their own yogurt using one of those "European" yogurt makers—basically an electric warming tray containing several small receptacles for holding the milk. They're still sold, but they're expensive, and you may have everything you need to make yogurt already at home.

Some interesting new science shows that rather than seeding a healthy intestinal ecosystem with new bacteria, yogurt culture actually enhances the ability of already established microbes to break down complex carbohydrates found in fruits and vegetables into nutrients capable of being absorbed by the intestines. In other words, you get more nutrients from the food you eat without having to eat more food because the yogurt enhances your intestinal flora's power to deconstruct the tissues of fruits and vegetables. The yogurt culture works hand in hand to boost the power of the ecosystem that's there, and then passes fairly quickly out of the body. The study was done by a team at Washington University in St. Louis led by Jeffrey I. Gordon and reported in *Science Translational Medicine*.

Yogurt culture is the same mix of microbes used to make cultured butter, namely Lactobacillus delbrueckii *subsp.* bulgaricus, Streptococcus salivarius *subsp.* bulgaricus, Lactobacillus acidophilus, and strains of bifidobacteria. You can buy these live cultures at www.culturesforhealth.com or at just about any market's yogurt department by buying a half pint of plain yogurt. Look at the ingredients. It should contain no coloring, texturizing, or thickening agents, just the milk and microbial culture. If it's organic, that's best, because then you're sure that there's no bovine growth hormone, antibiotics, or agricultural chemicals used in its production, and that the cows aren't fed GMO feed.

For equipment, you'll need a large stockpot and a smaller 4- or 5-quart stainless-steel pot with a lid. This smaller pot should be able to fit all the way into the larger pot. You'll also need a food thermometer with a clip so you can attach it to the side of the smaller pot, and a heating pad. Half-pint canning jars with lids and bands are perfect for the actual fermentation of the milk into yogurt.

Because you want the yogurt culture to do the fermenting, and not airborne lactobacilli and other free-floating microbes (if you want real yogurt instead of curdled milk), the milk will have to be heated to 185°F. This will kill any bacteria that may already be in the milk and clear the way for the yogurt culture—hence the need for the thermometer. It's also best to sterilize your pots by boiling water in them. Add the small pot's lid, the thermometer, a large metal spoon, a jar holder or tongs, and the half-pint canning jars, lids, and bands to the boiling water in the large pot. Ten minutes will kill any bacteria on the surface of the equipment. Again, we want just the yogurt culture to ferment the milk, not interlopers from the air. Remove the equipment from the boiling water, dry it quickly with paper towels, and cover it with a clean dish towel.

I like yogurt made with whole milk, but reduced-fat and fat-free milk will also make yogurt. It may not have the creamy smooth texture of yogurt made with whole milk, however.

Makes 8 half-pints

2 quarts whole milk, preferably organic

3 tablespoons organic plain yogurt with live cultures, at room temperature

1. Empty the small pot (see above). Pour enough water out of the large pot so that when the small pot is put into it, the water reaches about halfway up the sides of the small pot. The small pot will want to float, so press it down so it touches the bottom of the large pot. Add the milk to the small pot. Its level in the small pot should be about equal to the level of the water on the outside of the small pot. Return the large pot with the small pot in it to the stove over medium heat. Clip the thermometer to the side of the small pot so the bulb end is in the milk and you can read the temperature.

2. Keep the water in the large pot boiling and stir the milk frequently to keep the temperature evenly distributed. Meanwhile, put a stopper in your kitchen sink and add enough cold water so that it will reach about halfway up the small pot when you cool it later in the process. Add a couple of trays of ice cubes to the cold water.

3. When the milk reaches 185°F, it will begin to look frothy. Don't let it boil or overheat or you will ruin the milk for making yogurt. As soon as 185°F is reached, transfer the small pot to the sink. Don't let the cold water go higher than halfway up the sides of the small pot. Turn off the heat under the large pot. Stir the milk occasionally and continue to read the temperature. When it cools to 110°F, add the 3 tablespoons yogurt and stir well to incorporate. Put on the lid. Remove the small pot from the cold water.

4. Place a heating pad on a large cutting board and plug it in. Set the heat to medium. Dry the bottom of the small pot with a dish towel and set it on the heating pad. Cover the small pot and lid with a dish towel. (If you have a gas oven, the pilot light might give enough warmth. A temperature of 100°F is optimal. Check it with an oven thermometer. Few electric ovens have a setting as low as 100°F.)

5. Wait 7 hours without inspecting or stirring or in any way disturbing the milk.

6. At 7 hours, remove the towel and lid and taste the young yogurt. This is your baseline for taste. If it's as tangy as you like yogurt, then you're home free. If it's too tangy, knock an hour off the wait time for your next batch. If not tangy enough, let it go for 8 hours next time. It gets sourer the longer it sits.

7. You'll see that the milk has curdled, it has a pleasant cheesy aroma, and there may be some greenish whey on top. This is all normal. Use a clean spatula to stir the curds and whey vigorously so they are well incorporated into any remaining liquid. If the yogurt seems thin and runny, that's normal at this stage. It will thicken up in the fridge.

8. Pour the yogurt into the half-pint jars, leaving ½ inch head space. You can also use clean plastic yogurt containers that you've saved, or any other glass, ceramic, or food-grade plastic containers.

9. Put the lids on your containers and place them in the coldest part of the fridge. Chill overnight. The next day, they'll be the yogurt you're familiar with—only better. Add fresh fruit or use it plain on granola or in any other way you like yogurt. Your batch will last for 2 to 3 weeks. Be sure to reserve 3 tablespoons of this batch to start your next batch.

Greek Yogurt

Greek yogurt is simply regular yogurt that has been drained to remove most of the whey. It results in a thick, protein-rich yogurt that has become popular on its own and is used in many dishes the way sour cream is used, as in Greek specialties like tzatziki and Indian dishes such as raita and lassi.

Figure that you'll lose half the volume of regular yogurt when making Greek yogurt, and maybe a little more than half if using fat-free milk. So if you're starting with a batch of yogurt using the 2-quart recipe above, you'll end up with 1 quart (2 pints or 4 cups) of Greek yogurt.

- Have on hand: fresh, clean cheesecloth; a colander; butcher's string; a medium bowl. Make sure the colander fits in the bowl so it doesn't touch the bottom.
- Line the bowl with 2 or 3 or even 4 layers of cheesecloth. Dump the yogurt into the cheesecloth and then draw up the 4 corners of the cloth and hold it over the sink. Twist the corners to put pressure on the bag, forcing out as much liquid as you can.
- When most of the liquid has drained, it will drip more slowly. Now tie off the gathered ends just above the mass of yogurt with string. You can put this in the colander in the bowl to continue draining. Some people dispense with the bowl and colander and simply hang the cheesecloth bag from the faucet in the kitchen sink.
- Let the bag drain for 2 to 3 hours, then place the bag in the sink. Don't remove the string. Using your palms, press the bag to remove remaining liquid. When you've pretty much got it all, open the bag and with a spatula, turn the yogurt—now about the consistency of sour cream—into a clean storage container, cover it, and place it in the fridge. Let it rest 4 hours or overnight before using, and use it up within 4 to 5 days. Three tablespoons of this makes a good starter for your next batch of yogurt.

Mayonnaise Made with Whey

It certainly is easier buying a jar of Hellmann's or Best Foods at the market; but you really don't want that conventional food with all its additives, do you? Especially when mayonnaise is so simple to make at home, and so charged with beneficial bacteria when you make it with whey. If you already make kefir or yogurt, simply drain them through cheesecloth to

get the liquid whey. Most mustard is made with vinegar, giving mayo a bit of tang. Here we add whey for its sour nuance and its beneficial bacteria.

Makes about 2½ cups

> **3 egg yolks, at room temperature**
> **1 tablespoon freshly squeezed lemon juice**
> **½ teaspoon sea salt**
> **¼ teaspoon American yellow mustard**
> **1½ cups extra-virgin olive oil**
> **3 tablespoons whey**

1. In a bowl, whisk the yolks until they turn a light lemon yellow color.

2. Add the lemon juice, salt, and mustard and whisk until well incorporated.

3. Transfer the yolk mixture into a blender. Run the blender on medium-low speed and add the oil a few drops at a time to begin with. Adding too much oil too quickly yields thin mayonnaise. When the mixture thickens to the consistency of heavy cream, add the remainder of the oil in a very slow, thin stream. Easy does it.

4. When the consistency is that of mayonnaise, add the whey and blend it in thoroughly.

5. Transfer the mayonnaise to a jar with a lid and set it on the kitchen counter for 3 hours, then refrigerate. Use within 2 to 3 weeks.

Making Your Own Organic Cheese

Cheesemaking is fun and easy—if you have good milk to start with. The quality of the raw material is crucially important. You can make cheese from pasteurized and homogenized store milk, but it will not make as flavorful and interesting a cheese as will raw milk fresh from the animal. On the other hand, if you are curious about cheesemaking and want to give it a try, you may find it cheaper and easier to practice with store milk until you feel you have learned the essentials. And, because you made it yourself, it won't be bad at all.

Once you've learned the how-to, get in contact with suppliers of organic milk in your region. Cow's milk will be the easiest to find, goat's milk less so, and sheep's milk is at such a premium with cheesemakers that it may be hard to find any for sale. Of course, the best way to have a supply of raw, organic milk at your disposal is to have your own animals. That may or may not be possible for you, but in any event, keeping milch animals represents a real commitment of time and energy. There are no two-week vacations for folks who have to milk the goats twice a day, seven days a week.

To make a fine cheese, you'll need a source of starter culture, coagulant, and depending on the type of cheese you want, mold cultures, aroma-developing cultures, annatto color, lipase flavor enzymes, or other materials. Fortunately, all these products are available from Dairy Connection, Inc., 8616 Fairway Place #101, Middleton, WI 53562. I suggest writing for their catalog, which explains which cultures are used to make the various types of cheese. You can also peruse their products online at www.dairyconnection.com.

Rather than starting with the fancy equipment that cheesemakers use—perforated plastic draining baskets and so forth—we'll walk through the steps of simple cheesemaking using ordinary kitchen equipment. If you develop a passion for the craft, there will be time enough to move up to more expensive and dedicated equipment and materials.

First, make sure all your equipment is thoroughly clean. Use stainless-steel or food-grade plastic containers, such as those Rubbermaid makes. Have a supply of cheesecloth on hand. You'll need what's called a mesophilic starter culture that contains *Lactococcus lactis* subsp. *lactis* and *Lactococcus lactis* subsp. *cremoris*. Dairy Connection sells 20 units minimum (enough for 1,000 pounds of milk) for about $6. Rennet tablets for small batches are available at most supermarkets, usually the Junket brand, or you can buy coagulants through Dairy Connection. The starter culture suffuses the milk with bacteria that will ripen the cheese after it's made—its function is not to make curds. The coagulant causes the milk to make curds. Starter is sold as units, but people who don't want to get that technical can use the alternatives I list below.

Let's start with a small batch—say 2 gallons of milk—of what we hope will become something like a Colby or mild cheddar. Your milk should be at room temperature—ideally about 70°F. Add about a tenth of a unit of starter culture for 2 gallons of milk. In lieu of starter culture, you can practice your cheesemaking with buttermilk (½ cup for 2 gallons of milk), or with *Lactobacillus acidophilus,* available in liquid culture form at any health food or organic food store. Acidophilus won't make the kind of sincere little artisanal cheeses you're really after, but it will add some acid and help make a decent cheese. Use 1 cup of yogurt for 2 gallons of milk, or use 2 gallons of acidophilus milk to begin with. Allow the milk and culture mixture to stand at room temperature overnight.

Now add the coagulant. If you use rennet tablets, follow the directions on the package, but it's usually half a tablet for 2 gallons of milk. If you use lemon juice, it will take longer to coagulate and the curd won't be as firm. Don't go to the expense of chymosin coagulant until you're making large batches of cheese. The milk should be warmed to between 75 and 100°F during the coagulating period—closer to 100°F is better, but no higher. You can set your cheesemaking bowl into a larger bowl of warm water to maintain the temperature. After half an hour to an hour, test the milk to see if the curd has set and you get a clean break. Insert a finger down into the curd and lift. The curd should part cleanly and clear whey should appear in the break. The set may take up to 2 hours. This whey can be used within a week or two if kept in the fridge or up to 6 months if frozen. It's the required ingredient in whey mayonnaise (pages 178–79) and ricotta (pages 184–85), as well as many other probiotic recipes.

Once the curd has set, you'll need to cut it. Large pieces of curd will make a soft cheese, medium-size pieces (cubes roughly an inch or two on a side) will make a semi-soft cheese, and small curd (cubes ½ inch or less on a side) will make a firm cheese. Since we're after something like a Colby or cheddar, use a medium-size curd cut to cubes about an inch on a side. Don't worry if the curd isn't exactly the right size, or even if it granulates into something resembling cottage cheese. You'll still get cheese—it just may not be exactly the type or texture you hoped for. Use a sharp knife and cut the curd into inch-wide strips, then cut it crosswise into inch-wide cubes. If the curd is deeper in the bowl than an inch or two, turn the knife sideways and cut in parallel swipes from side to side. It will break apart somewhat, but don't worry. Keep the curd warm—to about blood temperature for a semi-hard cheese such as we're making.

After you've cut the curd, run your fingers through it gently for about 10 to 12 minutes, moving the curd pieces around until they firm up to the consistency of firm scrambled eggs. Don't squeeze the curd. Just stir it gently with your fingers. Make a bag over the sink with a couple of layers of cheesecloth laid into a mixing bowl and pour the curds and whey into the cloth bag. Tie up the ends of the bag and hang it from the faucet of the kitchen sink, allowing the whey to drain into the bowl. You may want to keep it and make a little ricotta. Allow the bag to drain naturally for about 6 hours.

While the bag is draining, prepare a cheese mold—a number 10 tin can used for a food product with both ends cut out is fine for this size batch. Use one of the cut ends (watch out you don't cut yourself) or find a small saucer or other flat ceramic circular material that's hard and won't impart anything to the cheese, and that just fits into the can. Place the can on a dinner plate and put the cheesecloth bag into the can, gently pressing it into the bottom so that it forms a round cylindrical cake. Press gently. Fold the excess cheesecloth over the top of the cake and put the saucer or can lid on top. Place a weight on the saucer—nothing too heavy, just something to exert some pressure so any excess whey will drain out. Something weighing a pound or so (approximately the weight of the cheese itself) is perfect.

Slide something under one end of the dinner plate so that any whey that oozes from the cheese will run off.

After a half hour, take out the weight and the saucer and turn the can upside down, sliding the cheese in its cloth to the opposite end, replace the saucer on top, and put on the weight. Repeat this 4 times over 2 hours. At this point, allow the cheese to sit under the weight for about 12 hours. At the end of that time, unwrap the cloth gently,

BOTTLE TO BE USED AS WEIGHT TO PRESS THE CHEESE

#10 CAN WITH ENDS REMOVED

CHEESE WRAPPED IN CHEESECLOTH

DINNER PLATE

and you should have a nice round cylinder of cheese. It's time for surface salting. First, rub the cheese all over with coarse salt (sea salt or kosher salt, not iodized salt) and allow it to mature for 2 to 3 days at a cool 60°F in a cool room or basement. Then rinse it in clean water and immerse the cheese in brine made by dissolving 12 ounces (1½ cups) of coarse sea salt or kosher salt (not iodized salt) in a quart of lukewarm water. Place the cheese in the brine so it's covered with liquid and allow it to stay there for 24 hours in the cool place you selected for maturing it.

After the 24 hours is up, remove it from the brine and place it on a tilted plate to dry. When the top is relatively dry, turn it upside down so the bottom can dry. By gently brushing (with a clean, damp cloth) and washing the surface of the cheese with brine several times during its maturation, you'll achieve a washed-rind cheese that prevents the development of stray molds on its surface.

Cheesemakers do sometimes add white rind mold or blue molds to their cheeses, or spray molds onto their surfaces, but that's a rather advanced technique. An alternative to the washed-rind cheese is dry salting. You can dry salt the cheese by gently rubbing the surface with kosher or coarse sea salt several times during the maturation period. Another alternative is to wait until a dry, yellow rind develops after about 2 weeks, then dip the entire cheese in melted wax. I don't particularly like this method, though, since I think the cheese needs to breathe during its maturation, allowing gases to escape and allowing air to support the natural microorganisms on the rind.

Maturation should take place in a cool (45 to 60°F) place with relatively high humidity. A root cellar or cool cellar is ideal. Think about a wine cellar—always at 58°F. There's a correspondence between cheese and wine; the ideal temperature and humidity (about 80 percent) for storing wine is also the ideal temperature and humidity for maturing cheese. Turn the cheese upside down weekly during its maturation. If you used raw milk, allow the cheese to mature for a minimum of 60 days, which ensures its safety, and longer if you want a sharper-flavored cheese. If you used pasteurized milk, you can allow it to mature for anywhere from 3 weeks for a very mild cheese to a year for a very sharp and tangy cheese. If some mold does develop on the dry rind, don't worry too much about it. Just brush it lightly and rewash the rind in brine, or brush it and dry salt it again. If some questionable mold persists, you can always trim it off when the time comes to enjoy your home-made cheese.

Making Your Own Ricotta

Ricotta is a by-product of the cheesemaking process. If you followed the recipe for homemade cheese above, you will have a couple of quarts or so of whey, which will make about a pint of ricotta—an Italian word meaning "recooked," because that's what you do to the whey to make the ricotta.

During the cheesemaking process, after you've drained the whey off the curds into a stainless-steel, ceramic, or plastic container (anything but aluminum, which will react with the acid), let it sit overnight. The next day, heat the whey until it is near boiling, about 180°F, but before it scorches, sticks, boils, or burns. You'll see it become cloudy with small white pieces of curd. Take the whey off the heat and allow it to cool until just warm, about 85 to 90°F, when it will be cloudy with fine curd and the liquid will be yellowish green.

Very carefully pour the clear whey off the curds into a couple of layers of clean cheesecloth lining a strainer suspended over a large bowl. Try not to disturb the curds. Here you're trying to get as much clear whey off the curds as possible without yet transferring the curds to the cheesecloth. You can discard the clear liquid or, if you have a pig, your pig will consider it a real treat.

Pour or scoop the curds from the cooking pot into the cloth. There will still be some whey mixed with them. Tie the ends of the cloth up into a bag and allow it to drain until all the whey is drained out. This may take several hours. You can give the bag a little shake now and then to help it to drain.

Scoop the well-drained curds into a container and either use fresh within two weeks or freeze indefinitely. The ricotta is unflavored. Add a little kosher or sea salt to taste.

Some people spice it up with lemon zest. You'll find that homemade ricotta spiked with lemon zest and a little egg yolk will make the best filling imaginable for rolled-up crepes, which you will then brown gently in butter. Serve hot with your choice of sour cream, crème fraîche, Greek yogurt, or fromage blanc. Oh, and strawberry jam. Yes, you're right. These are blintzes. Blindingly good blintzes.

Fromage Blanc

Fromage blanc is a thick and lovely fresh cheese, fermented, but not for long, yet containing the goodies produced by the bacilli in buttermilk. It's good over fruit, on crepes, and wherever you want a smoother, fresher, sour cream–like product. You can use the whey to make ricotta (page 184).

Makes about 1 generous quart

> **2 quarts whole milk**
> **1 cup half-and-half**
> **2 cups buttermilk**
> **2 tablespoons freshly squeezed lemon juice**
> **¼ teaspoon salt**

1. In a large saucepan, mix together the milk and half-and-half.

2. In a small bowl, combine the buttermilk, lemon juice, and salt, then add the mixture to the saucepan. Allow the mixture to rest for 1 hour, which gives the bacilli in the buttermilk a little time to work on the milk and half-and-half. (If you were to pour the lemon juice straight into the milk and half-and-half mixture, it could curdle the milk. It's best to incorporate it into the buttermilk and salt first because lemon juice doesn't curdle buttermilk.)

3. Gently, over low heat, begin raising the temperature of the liquid in the saucepan. Use a candy thermometer or food thermometer to measure the heat. Stir the mixture once or twice with a few quick strokes, as the heat rises.

4. When the temperature reaches 175°F, remove the saucepan from the heat and set aside for 10 to 12 minutes to cool.

5. Line a colander with several layers of cheesecloth and set it over a deep bowl. Pour the contents of the saucepan into the cheesecloth, then tie the ends up to form a bag. Hang the bag from the kitchen faucet to drain. When the draining stops, turn the fromage blanc into a container with a lid. If you wish, you can turn it into a bowl and beat it until very smooth if you prefer that texture. Cover the cheese and store in the fridge for up to 1 week.

Quark

On my first visit to Berlin, I was amazed at the ubiquity of ceramic stoves, oddly shaped toilets, incredibly good bread and beer, and quark—a lovely fresh cheese somewhere between ricotta and yogurt in texture and flavor. It's so easy to make, and so useful. Spoon some on granola and fresh fruit. Spread it on toasted bagels instead of cream cheese. Fill crepes, omelets, frittatas, and ravioli with it. Spoon it onto pasta. It's great stuff. It's also full of friendly bacteria.

Makes about 1 cup

> **1 pint whole milk, raw and organic if possible**
> **½ cup cultured buttermilk**

1. Pour the milk into a saucepan and heat to a simmer, then set the pan aside until the milk returns to room temperature. Stir in the buttermilk.

2. Pour the mixture into a glass or plastic container with a lid. Put on the lid and set on the kitchen counter or in a cupboard at room temperature for 24 hours, until it thickens to the consistency of yogurt.

3. Set a strainer in a bowl and line the strainer with cheesecloth. Pour in the thickened mixture and make sure the bottom of the strainer is far enough above the bottom of the bowl for the whey to drain off. Place it in the fridge overnight for the whey to completely drain off.

4. Use immediately or transfer the quark to a container with a lid and refrigerate. Use it within a week.

A Few Recipes Using Cheese

These recipes call for certain types of commercial cheese, but if you are making cheese at home, feel free to substitute your own. Aim to keep the texture and flavor of your homemade cheese as similar as possible to the commercial cheeses mentioned here. The easiest way to do this is to first use the commercial cheese recommended in the recipe, then match it as closely as practicable with one of your own. In a nutshell, the longer the cheese ages, the more intense its flavor. The more butterfat in the milk or cream, the softer the cheese will be, and softness depends also on the bacterial culture used. Blue cheese and many other highly flavored cheeses are made with special cultures. A good source online is www.dairyconnection.com, but there are other fine sources of cheese cultures.

Argentiera-Style Ragusano

Ragusano is an artisanal, organic Sicilian raw cow's milk cheese from pasture-fed, free-ranging cows, made only during the natural pasture season from November to May. The following recipe is from the Consorzio Ricerca Filiera Lattiero Casearia, which is working to protect Ragusano's natural heritage. Imagine this with your own cow's milk cheese. Wine recommendation: Stravecchio Siciliano from Vittoria, Province of Ragusa, Sicily.

Serves 4

¼ cup extra-virgin olive oil

1 clove garlic, crushed

14 ounces Ragusano Riserva cheese, cut into ¼-inch slices

4 ounces dried olives, pitted

Red wine vinegar

Fresh oregano leaves

In a large cast-iron pan, gently heat the oil with the crushed garlic. As soon as the garlic softens, remove it and lay the cheese slices in the pan. Gently brown the cheese on both sides. Remove the slices to a plate and keep them warm. To the remaining oil in the pan, add the olives and heat them over medium heat, sprinkling in a few teaspoons of vinegar and allowing the liquid to gently reduce, about a minute or two. Place the cheese on individual warm plates, top with the olives and reduced vinegar-oil mixture, and sprinkle with oregano.

Blue Cheese Meatballs

Forget the fancy cuisines of the two coasts. In Iowa, they're still doing down-home American cooking. But when you use a cheese as special and delectable as Maytag Blue, the results are anything but ordinary. Blue cheese culture is available from Dairy Connection (www.dairyconnection.com) and other sources. If you have access to goat's milk, goat blue cheese is a heavenly alternative to cow's milk blue.

Serves 2 Iowa farmhands or 4 light eaters

1 pound extra-lean ground beef
1 large egg, lightly beaten
1 medium onion, finely diced
½ cup plain bread crumbs
4 ounces blue cheese
2 tablespoons extra-virgin olive oil
2 tablespoons all-purpose flour
2 cups beef stock
5 ounces fresh mushrooms, sliced

In a large bowl, thoroughly mix the beef, egg, and onion together. Mix in the bread crumbs. Make 4 large meatballs out of the mixture. Punch a hole in the center of each meatball and insert 1 ounce of blue cheese. Seal the meatballs so the cheese is encapsulated.

Heat the oil over medium heat in a cast-iron skillet and lightly brown the meatballs on all sides, about 3 or 4 minutes. Remove them from the pan. Add the flour to the skillet and mix until the flour absorbs the oil and any liquid. Return the meatballs to the skillet and add the stock and mushrooms. Bring to a simmer and cook over low heat for 1 hour, turning the meatballs occasionally. Serve over rice.

Cheesecake Using Fromage Blanc

Down in West Virginia, Greg and Verena Sava make a delicious organic and fermented fromage blanc that substitutes beautifully for butter, measure for measure, in many dessert recipes. Here's their cheesecake recipe. It's simple and astoundingly delicious. You'll find the recipe for your own easy-to-make fromage blanc on pages 185–86.

Crust

    1½ cups whole wheat flour

    ½ cup wheat germ

    ⅞ cup Sucanat (whole cane sugar)

    ¼ cup extra-virgin olive oil

    ¼ cup ice-cold water

Filling

    3 large eggs

    ¾ cup granulated sugar

    Zest of ½ lemon

    4 cups fromage blanc

- **MAKE THE CRUST:** Preheat the oven to 350°F.

Combine the flour, wheat germ, and Sucanat in a large bowl. Add the oil and water at once and mix well until the mixture is soft and moist. Press the dough evenly into a 10-inch pie pan.

- **MAKE THE FILLING:** In the bowl of an electric mixer, whip the eggs with the sugar and lemon zest until well blended. Add the fromage blanc and again mix until all is well blended. Pour the mixture into the pie crust and bake for 1 hour, or until the center of the cheesecake is still wobbly and soft. Turn off the heat and let it cool in the oven with the door closed for 5 to 6 hours. This prevents the cheesecake from cracking and makes it smooth, creamy, and light. Cover and store in the refrigerator. Allow the cheesecake to come to room temperature before serving.

Kentucky Spoon Bread with Goat Cheese and Country Ham

When I asked Judy Schad of Capriole Farmstead Goat Cheeses, the artisan goat cheese maker from Greenville, Indiana, for a recipe using her fresh goat cheese, she didn't hesitate to give me this one from Joe Castro, executive chef at Louisville's famous Camberley Brown Hotel. It's rich, so make this a special treat—maybe for a cold November morning when you want to practice for Thanksgiving.

Serves 4 to 6

> **2 leeks, white parts only, cleaned and thinly sliced**
> **2½ cups whole milk**
> **2 cups half-and-half**
> **1 tablespoon sugar**
> **1 cup yellow cornmeal**
> **½ cup all-purpose flour**
> **½ cup (1 stick) unsalted butter**
> **6 large eggs, separated**
> **¼ cup heavy cream**
> **½ cup crumbled fresh goat cheese**
> **¼ cup diced country ham**

1. Preheat the oven to 350°F and butter a 2-quart casserole dish.

2. Place the leeks in a large saucepan and add water to cover. Bring to a simmer over medium heat and simmer until very tender, about 10 minutes. Strain the leeks, squeeze out the excess liquid, and set aside.

3. Combine the milk and half-and-half in the saucepan, scald it over medium heat, and gradually whisk in the sugar, cornmeal, and flour. Stir until thick, smooth, and creamy, about 3 or 4 minutes. Remove the saucepan from the heat and stir in the butter until melted.

4. In a large bowl, beat the egg yolks with the cream until light and smooth and add to the cornmeal mixture. Lightly beat in the goat cheese, ham, and leeks.

5. In a separate large bowl, whip the eggs whites until soft peaks form and fold into the cornmeal mixture.

6. Pour into the prepared casserole and bake until the eggs are set, 45 to 55 minutes. Serve warm with a little sorghum on the side.

Savory Cheddar Polenta with Maple Poached Apples

This recipe is so good, it made my mouth water when I read it over for the first time. It was even more mouthwatering on the plate. It comes from the Grafton Village Cheese Company—the cheesemaking arm of a nonprofit foundation that uses its proceeds to support preservation of the quaint village of Grafton, Vermont, and other charities. So besides doing yourself a good turn when you make this dish, you'll be doing the beautiful state of Vermont a good turn, too.

Serves 4

> 2 medium cooking apples, such as Honeycrisp or Granny Smith
> 2 cups apple juice or cider
> ¾ cup maple syrup
> ½ lemon, thinly sliced
> ½ cinnamon stick
> 5 whole allspice berries
> 2 tablespoons brandy
> 4 cups water
> 1 teaspoon salt

1 cup dried polenta

1 cup (4 ounces) shredded Grafton Village Gold cheddar cheese, plus more for
   topping (optional)

1½ teaspoons fresh thyme, or ¾ teaspoon dried thyme

Mascarpone cheese for topping

¼ cup toasted walnut or pecan halves for garnish

Brown sugar for topping

1. Peel, halve, and core the apples. Place in a 2-quart saucepan with the apple juice, maple syrup, lemon slices, cinnamon stick, allspice, and brandy. Bring to a simmer over medium heat and simmer for 15 minutes, or until the sauce thickens slightly and the apples are tender. Remove from the heat; keep warm while you make the polenta.

2. Bring the water to a boil in a large saucepan over high heat and add the salt. Gradually whisk in the polenta, reduce the heat to medium-low, and cook, stirring constantly, for 30 to 35 minutes, until velvety and the polenta pulls away easily from the side of the pan. Stir in the cheese and thyme.

3. Spoon one-quarter of the polenta into each of 4 shallow bowls or ramekins. Top each with half a poached apple and one-quarter of the poaching juices, followed by a generous tablespoon of mascarpone cheese. Garnish with the toasted walnuts and extra grated cheddar if you like. Finish with a sprinkle of brown sugar on top.

Yankee Doodle Macaroni and Cheese

The Yankee part of this dish would be Shelburne Farms aged cheddar. Like Grafton Village, Shelburne Farms is part of a nonprofit Vermont organization. In this case, it teaches the wise use of our natural and agricultural resources. The farm's beautiful buildings and grounds host thousands of visitors annually, and its herd of Brown Swiss cows, raised on the farm, produce rich milk that makes one of the world's finest natural cheeses. It's good community policy to buy some Grafton Village and Shelburne Farms cheese and thus sup-

port their efforts. And it gives you a reason to visit beautiful Vermont, especially during fall foliage time.

Serves 6

¼ pound Shelburne Farms aged cheddar cheese, grated

3 tablespoons unsalted butter

⅓ cup milk

Salt

1½ pounds rigatoni

½ cup heavy cream

½ cup grated Parmigiano-Reggiano cheese

1. Preheat the oven to 425°F.

2. Start bringing a pot of salted water for the pasta to a boil.

3. Meanwhile, combine the cheddar cheese, butter, milk, and a pinch or two of salt in a saucepan large enough to accommodate the rigatoni. Slowly warm over low heat, using a wooden spoon to break apart the cheddar and help it melt until the sauce is dense and creamy in texture, then take it off the heat and set aside.

4. Add the pasta to the boiling water and cook according to the package directions.

5. About halfway through cooking the pasta, add the cream to the cheddar sauce and return it to medium-low heat, stirring and reducing it slightly. Drain the pasta and add it to the sauce.

6. Turn the mac and cheese into a casserole dish and sprinkle the Parmigiano-Reggiano cheese over the top. Bake for 7 to 8 minutes, until the top bubbles and browns slightly. Don't overbrown. Serve immediately.

John Ash's Green Chile and Cheese Rice

John Ash is one of America's premier chefs; he's devoted to using organic ingredients because of his commitment to the environment. His book From the Earth to the Table won the Book of the Year award from the International Association of Culinary Professionals. John calls for Jack cheese in this recipe. I tried it with Cowgirl Creamery's all-organic

Mount Tam cheese and it was superb. Jack, brick, or Colby would each be excellent. Best would be one of your own well-aged washed-rind cheeses, such as those made with the recipe on pages 179–84.

Serves 4

½ pound semi-firm cheese, cut into ½-inch cubes

Fresh cilantro leaves

2 large poblano chiles, roasted, seeded, and cut lengthwise
    into ½-inch strips

1 tablespoon extra-virgin olive oil

½ cup minced onion

1 tablespoon minced garlic

1 cup basmati or long-grain white rice

1 teaspoon dried oregano

½ teaspoon fennel seeds

1¾ cups rich chicken stock

1. Make little "presents" that the diners will discover in their rice by wrapping each cube of cheese with a few cilantro leaves held by a strip of the roasted chile. Set aside.

2. Heat the oil in a medium saucepan over medium heat. Add the onion and garlic and sauté until softened but not browned. Add the rice, oregano, and fennel seeds and sauté for 2 minutes more, stirring occasionally. Add the chicken stock and bring to a boil. Lower the heat, cover, and simmer for 15 minutes, or until the stock is mostly absorbed but still visible. Test to make sure the rice is just about done. It should be soft through with just a little resistance in the very center of the grain.

3. Gently poke the chile-wrapped cheese cubes into the rice in various spots. Cover and continue to cook for 5 minutes, or until all the stock is absorbed. Remove from the heat and let stand uncovered for 3 to 5 minutes before serving.

Goat Cheese and Green Onion Galette

Joanne Weir is a thoroughly organic chef (as she should be, having spent five years cooking at Chez Panisse) who hosts a PBS show, Weir Cooking in the Wine Country, *and has written a companion book,* Weir Cooking: Recipes from the Wine Country. *I asked Joanne to pick a recipe she particularly loves and pair it with a wine. She says, "This galette has been one of my all-time favorites for years. You'll see why! Serve it with a full-bodied Chardonnay." It calls for several kinds of cheeses, plus crème fraîche, some of which, I hope, will be yours.*

Serves 6

1½ cups all-purpose flour, chilled in the freezer for 1 hour

¼ teaspoon salt

9 tablespoons unsalted butter, cut into ½-inch pieces, chilled in the freezer
  for 1 hour

⅓ to ½ cup ice water

1 tablespoon extra-virgin olive oil

2 bunches green onions, white and green parts, thinly sliced

5 ounces fresh goat cheese

4 ounces ricotta cheese

¾ cup coarsely grated mozzarella cheese

¼ cup crème fraîche

¼ cup grated Parmigiano-Reggiano cheese

Salt and freshly ground black pepper

1. Preheat the oven to 375°F.

2. Combine the flour and salt on a cold work surface. With a pastry scraper, cut the butter into the flour until half is the size of peas and the other half is slightly larger. Make a well in the center and add half of the ice water. Push together with your fingertips and set aside any dough that holds together. Add the rest of the water and repeat. Form the mixture into a rough ball. Roll out the dough on a well-floured surface into a 14-inch diameter circle and trim up the edges. Place on a large baking sheet and place in the refrigerator.

3. Heat the oil in a large skillet over medium heat. Add the green onions and cook until softened, 4 to 5 minutes. Remove from the heat and let cool.

4. In a medium bowl, mix together the green onions, goat cheese, ricotta, mozzarella, crème fraîche, and Parmigiano-Reggiano. Season with salt (easy on the salt—you might not even need any—since the cheeses have lots of salt) and pepper.

5. Remove the pastry from the refrigerator. Spread the cheese mixture over the pastry, leaving a 2½-inch border around the edge uncovered. Fold the uncovered edge of the pastry over the cheese mixture, pleating it to make it fit. There will be an open hole in the center revealing the cheese mixture. Bake the galette for 35 to 40 minutes, until golden brown. Let cool for 5 minutes, then slide the galette off the pan and onto a serving plate. It may be served hot, warm, or at room temperature.

CHAPTER 12

The Vegetable Ferments

Making Your Own Kimchi

In Korea, the traditional process of making and storing kimchi is called *gimjang*. In the era before refrigeration, and as far back as 1000 BC, Koreans prepared for the winter by filling large earthenware jars with chopped, brined Chinese cabbage, plus carrots, garlic, onions, turnips, leeks, daikon radishes, scallions, and other vegetables, mixed with a paste made with dried shrimp, fish sauce, ginger, and—after peppers from the Americas were imported in the sixteenth century—cayenne pepper. The jars were buried up to their necks in the earth and covered with thick layers of straw so they wouldn't freeze. Fermentation started in the fall when temperatures were above freezing, but as the winter set in, the kimchi stopped fermenting, allowing it to be used as a staple food during the cold months. In the

spring it would start fermenting again and the winter kimchi soon would be used up. As spring and summer vegetables became available, fresh kimchi would be made, maturing in a week so that fresh batches were made as needed with no need for storage.

As kimchi goes through its seasonal phases, different bacteria predominate, resulting finally in a stable spicy pickle that will last in the fridge for a month or more. There's an analog to this process in nature. Imagine a virgin forest in the eastern United States, undisturbed since pre-Columbian times. Let's say you take a bulldozer to such a climax ecosystem, like the mixed hardwoods, evergreens, understory trees and shrubs, perennial plants, and annual forbs in a mid-Atlantic forest, and you scrape it clean of all plants. A climax ecosystem will not grow back. First will come the opportunistic sun-loving rough weeds like giant ragweed, greenbrier, and pigweed. Then the tough grasses and brambles will return. Within a few years you'll get sumac and sun-loving woody shrubs like sassafras. Eventually trees like tulip poplars, quaking aspen, moosewood, and ailanthus will grow, throwing shade so understory woody shrubs that prefer shade, like the dogwoods, will be able to grow. Eventually, over many years, the evergreens like hemlock, plus birches, hardwoods like oak and hickory, and ash trees, hophornbeam, and beech will come back, but it will take many centuries for all these denizens of the eastern forests to settle into a climax ecosystem—stable, renewing itself year after year, supporting the rest of the flora and fauna native to the region.

The same kind of phases, where a host of microbes take turns bringing the kimchi to delicious maturity, happens when you make this spicy Korean pickled cabbage at home, only at a much more rapid time scale: weeks instead of centuries. Cabbage is held under the surface of salty brine. Certain bacteria of the genus *Lactobacillus* that happen to be floating in the air start to ferment the cabbage. As they work over a few days, they make the brine more and more acidic, which prevents pathogenic microbes from growing and encourages more and different acid-loving bacilli. The mix of microorganisms keeps changing until the kimchi is ready to be refrigerated. These microbes are also found in the fermentation of other foods, such as yogurt and cheese, and they are very, very good for us, as they add to the diversity and strength of the ecology of our intestinal flora. They also make kimchi taste great.

But here's the thing: The mix of bacteria, yeasts, and other microbes floating in the air in my kitchen will most likely be different from the microbes floating in your kitchen. And so every batch of kimchi is unique to the place where it's made. When you make it yourself, you are allowing the particular mix of microbes in your house or apartment to produce a pure expression of your very own locality. The French call this *terroir* and prize wines that show a taste of a certain place.

In Korea, the bacillus that ferments kimchi is called *Lactobacillus kimchii*. This bacteria produces bacteriocins (see page 25), which are proteins that destroy disease-causing bacteria and yeasts, and so help keep the kimchi wholesome as they also support our immune systems.

I first made kimchi at a gathering of the Fermenters Club at a winery in Santa Rosa. Austin Durant, the founder of the club in San Diego, flew up to Sonoma County for the get-together, but it was Jennifer Harris, a bundle of energy and a passionate fermenter, who put the session together. A generous handful of interested folks got a fermentation pep talk from Austin, and then Jennifer set us to work. She had previously filled a crock with three heads of shredded cabbage (she wanted to use the typical Korean napa cabbage, but the organic farm was fresh out so she substituted regular ballhead cabbage) and poured in enough brine to cover. The brine wilted the cabbage and pulled some of its juices into the liquid to make it crunchier.

We were instructed to bring a cutting board and knives, and we sat at long tables with vegetables that needed cutting. Within a half hour, many hands made short work of chopping whole carrots, daikon radishes, turnips, garlic, scallions, and jalapeño chiles into thin rounds. I asked Jennifer how I should cut my carrots, and she said, "As thin as possible." To me that sounded like a challenge, and I cut rounds so thin they were translucent. The guy next to me was cutting carrots into rounds a few millimeters thick, but I suppressed the urge to lecture him on how to cut thin rounds, since, at that point, I wouldn't have known what I was talking about. When all the vegetables were chopped, Jennifer and Austin tipped the crock and poured off the brine into a waiting container, then mixed the chopped vegetables into the cabbage. The brine wasn't discarded. There would be use for it later on. You can make the shredded cabbage soak ahead of the final assembly of the kimchi if you wish, but it's not required. It does help soften the cabbage for the squeezing and crushing that's coming later, but the kimchi will still be fine without that step.

Now Jennifer made the paste. This is a mixture of a thumb of ginger, a handful of garlic cloves, an optional ½ cup dried shrimp (you can get these at a Southeast Asian market), a couple of tablespoons of ground cayenne, and equal parts of sugar, sea salt, and fish sauce. But not just any fish sauce. According to Jennifer and Austin, Red Boat Fish Sauce is best because its quality is high and it contains no preservatives that would shut down the fermentation. Visit www.redboatfishsauce.com and hit the "where to buy" button. I entered "Santa Rosa, California" in the search box and found that the sauce is sold at five stores within easy driving distance from here. This sauce of fermented anchovies is about the closest thing we have to garum these days. Although it sounds yucky, it's a wonderful condiment and helps make kimchi the addictive side dish that it is. If you have a few minutes, google "garum" and enjoy a step back in time when ancient Greeks and Romans adored this sauce made from the fermented entrails of fish.

These ingredients are placed in a food processor or blender and whizzed into a paste. This paste is then turned into the crock with the vegetables and thoroughly mixed in by crushing, massaging, and kneading the paste-vegetable mixture with strong hands for about five minutes. You can wear latex gloves if you wish, but if Italians and Portuguese can mash grapes with their feet to make their wine, we can use bare hands to squeeze and crush our kimchi, no? Just wash your hands well before starting and don't touch your eyes if you're including hot peppers. Jennifer and Austin then filled a number of quart-size canning jars with the mixture and added enough of the original brine back so that the mixture of vegetables was covered.

Jen pointed out that dry vegetables protruding above the liquid offer bad bacteria and yeasts a place to grab a toehold.

Each of the attendees at this meeting of the Fermenters Club got a quart jar of this kimchi to take home, with some instructions. As the lactobacilli that are naturally floating in the air colonize the kimchi and start the fermentation, they will produce carbon dioxide bubbles in the mixture. These need to be evacuated. Jennifer did it by removing the lid and plunging a slender knife into the mixture here and there, allowing the bubbles to rise to the surface and the gas to escape. She said that when we arrive home with our quart of kimchi, we should remove the lid, cover the top of the jar with a square of paper toweling, and screw on the band that holds the lid but without the lid itself. Gases will be able to escape through the paper, but fruit flies or other unwanted visitors won't be able to get in. And store the kimchi in a cupboard that has a moderate temperature—not too hot and not too cool, with the ideal about 75°F. Taste it after a week, and if it seems about right (a previous batch she made a week before tasted great) to your taste, put on the metal lid and band and store it in the fridge. But every day or two, open the jar not only to eat some of the kimchi, but to let any accumulated gases escape.

Kimchi

This is the recipe for kimchi that I suggest you start with. You can add and subtract vegetables as you see fit and as vegetables become available through the seasons. Just be aware that summertime kimchi will ferment rapidly, while cold winter kimchi will take more time to get just right. And remember, too, that refrigeration slows fermentation to a crawl, putting the microbes into a kind of suspended animation. This recipe can be doubled or tripled, depending on how many hungry kimchi recipients are waiting. Of course, all ingredients should be organic. Vegetables or other ingredients dosed with pesticides or preservatives will kill off or set back the beneficial fermentation microbes.

Also, in my home batches of kimchi, I sometimes use one head of green cabbage and one head of dark purple cabbage. When the fermentation is finished, the cabbage in the kimchi turns a beautiful pink color. Same holds true for sauerkraut. If you don't want pink kimchi or sauerkraut, don't use red cabbage.

A final word about this recipe. Follow it for your first batch, then relax and get creative with subsequent batches. I just made a batch using about 2 dozen rosettes of tatsoi, a small cabbage-family plant with spadelike leaves that we grew in our garden, plus a couple of our own garden carrots. Instead of cutting the carrots into rounds, I shredded them so they would ferment more completely. Instead of cutting a daikon radish into rounds, I peeled a root, cut it in half crosswise, then cut each 5-inch piece into long sticks, the way you'd cut a carrot when making celery and carrot crudités. I had 2 small heads of bok choy. One I reduced to whole leaves, the other I cut crosswise into ¼-inch slices. In other words, ferment what you like, cut it as you please, and keep the vegetables weighed down and under the brine for safe and tasty results.

Makes 3 or 4 pints

Vegetables

½ cup sea salt

2 quarts filtered or spring water

1 large head napa cabbage

3 medium carrots

1 daikon radish

3 scallions

Paste

3 serrano chiles, or to taste

2 tablespoons dried shrimp (optional)

One 5-inch piece fresh ginger, peeled and grated

6 cloves garlic, peeled and minced

½ cup sugar

½ cup fish sauce (without preservatives)

1. Place the salt and water in a ceramic crock or glass container and stir until dissolved. Remove the outer leaves from the napa cabbage and slice the remainder crosswise into ¼-inch slices. Place these in the crock with the brine.

2. Slice the carrots, daikon radish, and scallions into very thin rounds and mix them into the cabbage and brine. Place a plate on the vegetables to hold them under the brine. Weigh down the plate with a closed jar of liquid, bottle of wine, or gallon-size, closed zip-top freezer bag with at least a quart of water in it for 6 hours, either during the day or overnight.

3. After the soak, drain the vegetables in a colander and place them in a bowl. Reserve the liquid brine.

4. Stem the chiles and slice them in half lengthwise. Use as is if you want a hot and spicy kimchi, or for less heat, remove the seeds and membrane and discard. Mince the chiles and place them in a bowl.

5. Add the shrimp (if using), ginger, garlic, sugar, and fish sauce to the bowl with the chiles. Transfer to a food processor or blender and whiz to a thick paste.

6. Put half the vegetables and half the paste back into the original crock or jar and mix thoroughly. Add the remaining vegetables and paste and mix thoroughly again.

7. Crush the vegetables with your hands, as if performing deep tissue massage, squeezing and crushing, for about 5 minutes, until all the vegetables are thoroughly crunched.

8. Put a plate on the vegetables in the crock to push them slightly under the juices. If the top seems dry, add a little of the reserved brine to make sure everything is wet. Put a weight, such as a closed quart jar of water, on top of the plate to keep the vegetables under the surface.

9. Cover the crock containing the submerged ingredients with a cloth to keep out insects. Punch down the kimchi every day for a week to release carbon dioxide gases and to mix the ingredients. When it tastes right, about a week or two later, spoon the kimchi into canning jars, add a little of the brining liquid so everything stays wet, screw on the lids with metal bands, and store in the fridge for up to 3 or 4 weeks. When you open the jars to use some of the kimchi, you'll allow any buildup of CO_2 to escape. Be sure to share with family and friends.

Korean Mung Bean and
Kimchi Pancakes (Nokdu Bindaetteok)

Jennifer Harris, my original kimchi teacher, posted this recipe for her Facebook friends to try. It combines beans, vegetables in the kimchi, and rice flour for a nonmeat source of good protein, although ½ cup of cooked ground pork could be added to give the pancakes a flavor boost. The dipping sauce is powerfully salty, so you might consider making a sweet and sour sauce (equal parts simple syrup and rice vinegar with a squeeze of lime juice) instead. Unless the kimchi has some strong spicy heat from chiles, the pancakes will be rather bland. You can spoon Sriracha or other hot sauce on the pancakes to enliven them, but if the kimchi is spicy, it will make a more integrated flavor and overcome the blandness. Don't overcook to dark brown—these are best when golden brown. And they're filling. One pancake was plenty for me. Finally, they are really good.

Makes four 6-inch pancakes

Batter

1 cup split dry mung beans
1⅔ cups water
⅓ to ½ cup rice flour
¾ cup kimchi (pages 201–3)
Cooking oil, such as canola

Dipping sauce

¼ cup organic tamari
2 tablespoons rice vinegar
½ teaspoon sesame seeds

1. Soak the mung beans in a bowl with the water overnight.
2. Combine all the dipping sauce ingredients in a bowl, cover, and refrigerate.
3. Drain the beans, reserving the soaking liquid. Place the beans in a blender or food processor with 3 tablespoons of the soaking liquid and whiz until they form a smooth paste.

4. Add ⅓ cup rice flour to the paste and blend again. If the batter seems too thin, add the rest of the rice flour. If it's too thick, add more soaking liquid. Aim for the texture of a typical pancake batter.

5. Fold the kimchi into the batter, mixing thoroughly.

6. Heat 1 tablespoon of oil in a small skillet over medium heat. If it smokes, it's much too hot. Add ½ cup of batter to the pan and cook for 2 to 3 minutes, until the rim of the pancake is a deep, golden brown, then flip and cook for 2 to 3 minutes on the other side. Repeat with the remaining batter to make 4 pancakes, adding more oil to the skillet as needed. Serve hot with the dipping sauce.

Making Pickles

When I lived in eastern Pennsylvania among the Pennsylvania Dutch (who were Deutsch—German—and not Dutch), I shopped every week at their markets, especially at Zerns in Gilbertsville and Renningers in Kutztown. These were the real thing, with Amish and Mennonite horse-drawn buggies tied up outside, and true Pennsylvania Dutch farmers offering their foods for sale inside.

Among their sausages, scrapple, and head cheese, they also offered chowchow—a pickle of many kinds of summer vegetables. What I realized about all these foods was that they originated many centuries ago, when the progenitors of the Amish and Mennonites were still in the Black Forest of Germany, when there was no refrigeration, and when this was how you preserved foods for the winter months. The fact that pickled vegetables bring a load of lactobacilli to the diet was a bonus.

The fabled "pickle barrel" of the general store of yesteryear also contained brined cucumbers preserved by lactobacilli. Even today, when you mention pickles, most people think first of cucumbers, although just about any vegetable can be pickled, as you can see in the pickled vegetable medley that is chowchow, usually consisting of green tomato, cabbage, chayote, red tomatoes, onions, carrots, beans, asparagus, cauliflower, and peas. The Pennsylvania version is usually made sweet with added sugar, while the version served across the Deep South came to Louisiana when the English forced the French settlers called Acadians out of Nova Scotia.

These refugees found a home in French-influenced Louisiana, where they are known to this day as Cajuns. Cajun chowchow is less sweet than the Pennsylvania version.

Because we are a hamburger-eating country, and a sliced dill pickle goes so well with that and many other sandwiches, let's start making pickles with cucumbers and then move on to chowchow. The fermented cukes I've made are nothing short of wonderful. I serve them to visitors to my home, and they say they're the best pickles they've ever tasted. So be prepared for high quality.

As spring fades into summer, cukes will start appearing at the farmers' markets. Look for pickling cukes about 3½ to 5 inches long and about the width of a plump sausage, rather than big, long ones or short, round ones. If you want to make small cornichons, look for 2- to 3-inch warty, curved cukes. If you plan on growing your own, there are superior varieties to be found at www.rareseeds.com. Look for thin-skinned varieties bred specifically for pickling, and harvest them while they are still young and 5 inches in length or less. As they grow bigger and longer, the seeds inside start to mature and turn hard, making them less pleasant to eat.

Once again, it's the lactobacilli that work their wonders on cucumbers, turning them into pickles by metabolizing the vegetables' sugars into lactic acid, imparting flavor, enhancing texture, and helping to preserve them—to say nothing of their beneficial probiotic effects in your digestive system. Even when you can excess pickles in canning jars to store them, killing the lactobacilli, you still get the benefit of the metabolites they produced when alive.

If there's anything you want most in a pickle, it's a snappy crunch. And the brine helps with that. Consider: The water inside a cucumber has a much higher water-to-salt concentration than the brine, which has a higher salt-to-water concentration. Nature likes to even things out, and so through a process called osmosis, water in the cucumber flows out through the membrane of the cucumber skin to try to equalize the concentration, rendering the pickle crunchier than when it was first put into the brine.

Best darn pickle ever!

Garlic Dill Pickles

This recipe can be scaled up or down, but I wanted to give you the basic instructions for 10 pounds of cukes. As for equipment, have a vegetable brush handy. You'll need a 5-gallon ceramic or glass crock or a 5-gallon bucket of food-grade plastic. Please make sure it's food-grade; that is, it originally was used to hold food. Other plastics leach toxic chemicals into their contents. You'll need a glass or ceramic plate that just fits inside the crock or bucket, plus an unused gallon-size zip-top freezer bag, clean dish towels, a fresh package of cheese-cloth, a large stainless-steel or other nonreactive metal pot, a carton of canning jars with lids and bands, and a narrow plastic spatula.

Note that the recipe calls for pickling spices. You can find them in the spice rack at almost any supermarket, but if you want to make your own, mix together crushed cinnamon sticks; bay leaves; ground allspice, mace, and ginger; the whole seeds of mustard, dill, black peppercorns, coriander, juniper berries, and cardamom; plus whole cloves.

Makes 3 or 4 quart jars, depending on the size of the cukes

> **10 pounds unwaxed pickling cucumbers**
> **¼ cup pickling spices**
> **2 bunches fresh dill**
> **1 cup white vinegar**
> **1 gallon spring or filtered water**
> **¾ cup coarse pickling salt (not iodized) or sea salt**
> **10 cloves garlic, peeled**

1. Scrub the surface of the cucumbers under cool running water to remove any incidental soil. Cut ¹⁄₁₆ inch off the blossom end of the cukes. Blossom ends contain enzymes that can render your pickles soft. Cut the stem ends back to where the cucumber flesh starts. Discard any cukes that are discolored, bruised, or soft.

2. Put half the pickling spices and 1 bunch of dill in the bottom of your fermenting vessel. Add all the cucumbers.

3. Mix the vinegar and water in a large bowl. Add the pickling salt and stir to dissolve it completely. Pour the mixture over the cucumbers. Add the garlic, the rest of the pickling spices, and the second bunch of dill.

4. The cucumbers must be fully submerged under the brine at all times during the fermentation. Use a glass or ceramic plate that just fits in the vessel to weigh them down. Fill the gallon freezer bag with more vinegar-salt-water brine, shut it tightly, and set it on the plate (if it leaks, no harm done, as it won't dilute the brine in the vessel; you may need some for topping up the vessel, and you may need it later if you are canning some of the pickles).

5. Cover the vessel with a clean dish towel and place the crock in a spot where a temperature of from 70 to 75°F is maintained. Lactobacilli work best at this temperature. Lower or higher temperatures favor unwanted spoilage bacteria or fungus spores.

CANNING PICKLES

To can pickles, place them upright in wide-mouth canning jars that have been boiled along with their lids and bands to sterilize them. Fill the jars with brine brought boiling from the stove, leaving ½-inch head space in the jars. Insert the spatula and work it up and down around the pickles to remove trapped air. Place a lid and band on each jar and screw it down just until you feel a little resistance—finger tight. Process in a boiling water canner for 15 minutes. If you don't have a canner, use a large pot with a rack on the bottom to keep the jars from touching the bottom of the pot. Add enough water to reach from one third to halfway up the sides of the jars and boil covered for 15 minutes. Remove the jars from the canner or pot and give the bands an extra twist to tighten. The lids should soon pop down, showing that they have sealed. Store the pickles on a cool, dark pantry or basement shelf. Wait at least 1 week before using, to give the pickles a chance to integrate their flavors.

Additions to the Pickle Crock

When making your pickles, add 2 fresh grape, cherry, oak, or bay leaves to the crock. These will give the pickles a little extra stringency from the tannins in the leaves, and a slightly different new flavor.

6. Check the crock every day, but don't taste the pickles. After a day or two, you'll see some scum forming on the surface of the brine. This is yeast growth and must be removed or the pickles will spoil. Remove it every day. Keep the brine topped up with extra brine from the bag if needed.

7. Let the cucumbers ferment until they become an even olive green color, about 2½ to 3 weeks. Taste a pickle. If it has good dill flavor and a sour taste, they're done. If you want more time, allow them to continue fermenting, but no longer than 3 weeks. Pour off the brine into the stainless-steel or other nonreactive pot through several layers of cheesecloth to remove the solids and impurities. For immediate consumption, up to 6 to 8 weeks, store the pickles in the fridge in jars topped up with the brine. For long-term storage, you'll have to can your pickles.

Chowchow

Many people preserve chowchow by covering the vegetable medley with a vinegar, sugar, and salt mixture and processing it in a boiling water bath for 20 to 25 minutes. But chowchow can also be fermented, following the same regimen used to ferment pickles. Instead of cucumbers, make a medley of the following chopped vegetables, adding others or subtracting to your personal preference.

Makes 3 to 4 quarts

> 5 cups coarsely chopped green tomatoes
> 5 cups coarsely chopped ballhead cabbage
> 1½ cups chopped yellow onion
> 2 cups coarsely chopped red and green bell peppers
> 2½ cups cider vinegar
> 1 cup packed light brown sugar
> 1 tablespoon yellow mustard seeds
> 2 cloves garlic, minced
> 1 teaspoon celery seed
> ½ teaspoon crushed red pepper, or to taste

1. Prepare and place all ingredients in a crock or food-grade plastic container.

2. Make enough brine (1 cup non-iodized or pickling salt to each quart of water) to cover the vegetables in the crock. Fit a plate on top of the vegetables so that it is submerged under the brine. Weight it with a quart jar filled with water and sealed.

3. Cover the crock with a dish towel and inspect it every day or two, removing any scum that floats to the surface.

4. After 3 weeks, pour off the brine through a strainer and reserve. Spoon chow-chow into quart jars with lids and top up with the strained brine. Fix on the lids and store in the fridge. To reduce saltiness, place chowchow for that day's use in a strainer and rinse it with cold water.

Fermented Pickle Relish

If you're making those fabulous pickles from the recipe on pages 207–9, you could prepare a pickle relish by mincing several of those pickles and adding some finely chopped onion— or you can make this recipe. Load your next hot dog with mustard, your own sauerkraut (pages 215–17), and this pickle relish. Or make a sandwich with sausage, grilled onions and peppers, chili sauce, sauerkraut, and this relish. Yum-ola.

Makes 1 quart

> **6 pickling cucumbers, minced**
> **1 small yellow onion, chopped**
> **3 tablespoons chopped fresh dill**
> **1½ tablespoons sea salt**
> **5 tablespoons fresh whey drained from yogurt**

1. Place the minced cukes in a medium bowl and add the remaining ingredients. Mix to incorporate well and make sure the salt dissolves.

2. Place the mixture in a canning jar and press the mass down hard so the juices run. The liquid should cover the cucumber mixture. If it doesn't, add enough filtered water or spring water so that the relish is covered.

3. Lay a piece of paper towel over the jar mouth and screw down one of the canning bands.

4. Set the jar in a kitchen cupboard or on the counter for 2 days, then refrigerate. Use within 1 month.

MamaKai's Butternut Squash and Carrot Kraut

This recipe is from Angie Needels, director of MamaKai, a Berkeley business that offers ready-to-eat, probiotic-rich, nutritive-dense meals and snacks to growing families planning for their birth and postpartum. By having healthful meals on hand, these new mamas can better enjoy their newborns while maintaining vital health in caring for them. Visit and learn more at mamakai.org.

The quality of your ingredients is important in this and all recipes; for more nutrients in your produce, try to buy organic and from your local farmers' market when possible.

Makes about 2 quarts

> **4 cups filtered or spring water**
> **4 teaspoons sea salt**
> **1 large butternut squash, peeled, seeded, and thinly sliced with a mandoline**
> **1 large leek, white part only, thinly sliced with a mandoline**
> **½ medium daikon radish, cut in half and thinly sliced with a mandoline**
> **1 bunch carrots, grated**
> **2 tablespoons grated fresh horseradish root**
> **6 cloves garlic, mashed in a garlic press**
> **2 tablespoons grated fresh unpeeled ginger**
> **½ bunch cilantro, finely chopped**
> **2 tablespoons dulse flakes**
> **1 teaspoon chili paste**

1. Combine the water and salt in a large bowl and stir well to dissolve, making the brine starter.

2. Fold the remaining ingredients together in a large bowl until they are well incorporated.

3. Ladle the ingredients into either 1 large glass or ceramic fermentation container or mason jars of desired size. Smash the ingredients down into the vessel(s) and fill with brine. Smash down once more to make sure that water has seeped into all nooks and crannies and all veggies are below the water solution. If using a large fermentation crock or container, cover the contents with a plate topped with a weight (a smaller mason jar filled with water can be a great weight). If just using quart- or pint-size mason jars to ferment in, no weight is needed, but you should keep checking to make sure that the vegetables are submerged under the brine at all times.

4. Cover and allow to ferment, opening and smashing down every 2 to 4 days or so. You'll notice that your ferments will be very bubbly and vibrant in the first week or two (that's how you know it's working), but will slow down after that.

5. Continue fermenting past this phase to allow the flavors to meld more, typically 4 to 6 weeks, and to get a very bright, sour, fermenty flavor. Stop the fermentation by placing the vessels in the fridge when the kraut tastes best to you.

Fermented Turnips

It couldn't be simpler to make fermented turnips, but the rewards are a complex of the health benefits of all cabbage family vegetables, of which turnips are one, and excellent flavors. Use small white Tokyo Cross turnips if you can find or grow them. The Cultured Pickle Shop in Berkeley ferments them in a brine with added turmeric.

Makes about 1 quart

> **2 pounds organic turnips, trimmed and well scrubbed**
> **1 tablespoon sea salt or pickling salt**

1. Grate the turnips into a large bowl.

2. Sprinkle the salt onto the grated turnips and toss to combine well.

3. Pack the grated, salted turnips into a wide-mouth quart canning jar and cover with filtered water, leaving enough space to make sure the turnips are kept submerged by putting a pint-size zip-top freezer bag filled with water on top of the brine.

4. Cover the jar top with a paper towel and a canning band and screw it down. Place it on a warm kitchen counter for 3 to 5 days, until bubbles are rising and visible.

5. Taste the fermented turnips. If they're to your liking, put a lid on the jar, screw it down tightly, and store in the fridge, finishing the turnips within 2 to 3 weeks.

Grate the turnips into a large bowl.

Daikon with Turmeric, Fenugreek, and Cardamom

Another goodie from the Cultured Pickle Shop.

Follow the directions for Fermented Turnips (page 212), substituting daikon for the turnips and adding 1 teaspoon ground turmeric and ½ teaspoon fenugreek seeds, plus 8 to 10 cardamom seeds husked out of the pods for each quart of ferment.

Curtido

Folks in southern Mexico and Central America like a bit of this fermented vegetable mix with every meal. It's the Latino version of coleslaw, sauerkraut, or kimchi, and it's eaten as a side dish. A Salvadoran pupusa wouldn't be complete if it weren't served with curtido.

Makes 2 to 3 quarts

1 head cabbage, cored and shredded

1 cup grated carrots

2 medium yellow onions, quartered and thinly sliced

1 tablespoon dried oregano or 2 tablespoons fresh oregano leaves

½ teaspoon crushed red pepper flakes

1 tablespoon sea salt or pickling salt

¼ cup whey (or substitute 1 more tablespoon salt)

1. Place all the ingredients except the whey in a large bowl or stainless-steel pot.

2. Pound with a wooden pounder for 10 minutes to release the vegetables' juices.

3. Place the vegetables in a wide-mouth container or crock. Add filtered water to just cover, then add the whey or the extra salt. Stir.

4. Weigh the vegetables down with a plate on which a weight, such as a half gallon of water, is set. The vegetables must be submerged and kept away from air. Cover the container with a dish towel.

5. Leave at room temperature for 3 to 5 days, removing any scum that rises to the surface of the liquid at each inspection. After the fermentation, place the curtido in the fridge, where it will keep for 2 to 3 months, improving with age. If the mixture turns gray or develops an off odor, discard it.

Gingered Carrots

There's something about the flavor of ginger that augments the flavor of garden-fresh carrots. If you're not growing your own, look for carrots with their tops on at the market. If the tops appear bright green, aromatic, and fresh, the carrots will be, too. The fermentation period helps these two disparate—but symbiotic—flavors to meld and mellow.

Makes 1 quart

4 cups tightly packed grated carrots

1 tablespoon grated peeled fresh ginger

2 teaspoons sea salt or pickling salt

3 tablespoons whey (or substitute 1 more teaspoon salt)

1. In a large bowl, mix all the ingredients together and set the bowl aside for 30 minutes.

2. Using a wooden pounder or potato masher, pound the carrot mixture for 5 minutes so the vegetables release their juices.

3. Place the contents of the bowl into a quart canning jar and press down firmly so the carrots are covered by juice. Add filtered water to just cover the carrots if needed. Leave 1 inch of headspace between the top of the juice and the top of the jar.

4. Place a single sheet of paper towel over the jar and screw on the band that holds the jar's metal lid.

5. Set the jar on a warm kitchen counter for 3 days if using whey or 5 days if using only salt. Check the jar each day and remove any scum that rises; keeping the vegetables submerged will prevent any spoilage. Then cover the jar with a metal jar lid before placing it in the fridge, where it will keep for 3 to 4 weeks.

Sauerkraut

Cabbage, like many of the cole crops (its cruciferous relatives in the plant world), grows well in cool weather, which is why it was grown extensively across northern Europe and the northern tier of the United States in the seventeenth through the nineteenth centuries, before refrigeration. Winter closes in fast in these cold regions, and hard freezes would destroy the cabbages that provided farmers with greens and vitamin C during the long, freezing months.

And so those farmers learned to do a simple trick—brine their cabbage until the vegetable was fermented and stabilized, whereupon the cabbage would last just fine in cold storage during the winter. When some was needed for the table, it was just fished out of the crock and the lid replaced. As we now know, cabbage is a wonderfully nutritious food, providing vitamin C and other essential nutrients, protecting the human body against diseases like cancer, and the lactobacilli that colonize the sauerkraut release bound-up nutritive factors in the cabbage—sauerkraut contains twenty times more bioavailable vitamin C than raw cabbage—and for the human gut.

Sauerkraut is a German word meaning "sour herb," and when the huge influx of

Germans came to America in the late seventeenth to the nineteenth centuries, they brought their sauerkraut—or the knowledge of how to make it—with them. Other Americans, too, have a long history with sauerkraut, not only in German enclaves like Cincinnati, Milwaukee, eastern Pennsylvania, and even Yorkville on Manhattan's Upper East Side, but among the general population, where the all-American hot dog is often loaded with kraut.

The beautiful part is that it always has been and continues to be a snap to make at home. This recipe is from Austin Durant of San Diego, founder of the Fermenters Club (www.fermentersclub.com). For your first batch, you may want to try it without any additions other than cabbage and brine and see how you like it.

Makes 4 quarts

5 pounds (2 or 3 heads) cabbage
3 tablespoons sea salt or pickling salt (non-iodized)
1 teaspoon caraway seeds (optional)
½ teaspoon juniper berries (optional)

1. Remove just the outer leaves of the cabbage and slice the heads in half through the core. Using a sharp knife or a mandoline, slice shreds off the core, but not the core itself. Discard the cores when finished. You can also remove the cores before slicing.

2. Place the shredded cabbage in a large bowl. As you add handfuls, sprinkle the cabbage with a little salt and the spices, if desired. When all the cabbage is shredded and in the bowl, add the remaining salt and spices.

3. Using your hands, toss, squeeze, pound, crunch, and massage the cabbage for about 10 minutes, until the shreds grow limp and the cabbage juices start to run. As the cabbage turns limp and juicy, use a wooden mallet or pounder to finish the cabbage for the last 3 or 4 minutes.

4. Place the cabbage in a small ceramic crock or wide-mouth 1-gallon glass jar. Place a plate that fits in the crock or jar on top of the cabbage and press down hard to work any air bubbles out of the cabbage.

5. Weigh down the plate with a gallon jug of water or with zip-top freezer bags filled with brine. (Freezer bags can leak, and if your bag does, then it will only add more brine rather than diluting the ferment with plain water.) Place the bag unsealed

on the plate with the seal up so it spreads to cover the plate and any juice showing between the plate and crock. Then seal it. The juice should completely cover the cabbage. If there's not enough juice to cover the cabbage, put a little extra brine in the crock or jar until the cabbage is entirely under the liquid.

6. Cover the crock or jar with a clean dish towel held in place with a rubber band and store it in a place that ideally is between 70 and 75°F.

7. Check the crock or jar after a couple of days and every few days after that. It will start to ferment (bubble) after a few days. If you see any mold growing on the surface of the liquid, remove the weight bags or jug and the plate, and skim off as much as you can. Don't worry, the sauerkraut is safe as long as it's submerged under the brine. Just get as much scum or mold off the liquid as you can and replace the plate, weight, and towel. The kraut will improve in flavor over the next month or two. If you find it exactly at the place you like it, pack the kraut into canning jars with brine to cover and store them in the fridge, where it will last for several months as long as the kraut is covered with brine. Rinse the kraut to remove salt before serving if you wish.

German Apples and Kraut

What goes with a big helping of roast pork? Applesauce, sauerkraut, and potato pancakes, especially if your heritage is German. Two of these constituents can be fermented together—with glorious results: apples and cabbage. Here's how.

Makes 2 quarts

> 1 *medium head ball cabbage, shredded*
> 1 *teaspoon sea salt or pickling salt (non-iodized)*
> 2 *firm apples, peeled, cored, and shredded*
> 1 *teaspoon grated fresh ginger*

1. In a large bowl, combine the shredded cabbage and salt. Work the salt through the cabbage, massaging the shreds for 5 to 10 minutes, until the cabbage juice runs. Add the shredded apples and ginger and then massage again for a minute to incorporate.

2. Pack the kraut into a small crock or glass or ceramic bowl and place a plate on top. Set a quart canning jar filled with water and with its lid screwed on as a weight on the plate. If there's not enough juice to cover the cabbage, add just a bit of water until the cabbage is covered. Cover the crock or bowl with a clean dish towel.

3. Place the kraut container in a warm place—70 to 75°F is ideal—for a week, checking daily to skim any foam from the top and removing the weight and plate and stirring the kraut 2 or 3 times during the week.

4. Strain the kraut, reserving the juices in a bowl.

5. Pack the apples and sauerkraut into quart glass canning jars, adding enough of the reserved juice to keep the kraut wet, and store in the fridge. The kraut lasts for 2 to 3 weeks.

Escabeche

Escabeche is a Spanish and Provençal dish of poached or fried fish in a spicy-hot marinade, usually served cold as an appetizer. Here we're forgoing the fish and making pickled peppers fermented by our friends the lactobacilli. It's meant to be fiery hot, used as a condiment on tacos, over meats or fish, on sandwiches, or whenever some real heat is needed to brighten up a dish. When working with fiery-hot peppers, it's a good idea to use clean plastic kitchen gloves and wash them off thoroughly when you're finished.

Makes about 2 quarts

4 cups thinly (¼ inch) sliced seeded hot and sweet peppers of your choice (see Note)

4 carrots, peeled and sliced on the diagonal into ¼-inch ovals

6 cloves garlic, peeled and chopped

½ medium onion, cut stem end to root end, then sliced into half rounds with the rings separated

1 tablespoon black peppercorns

1 teaspoon chopped fresh oregano

2 tablespoons sea salt or pickling salt

1. Sterilize 3 quart-size canning jars, lids, and bands in boiling water for 10 minutes.

2. Place all the ingredients in a large bowl. Mix thoroughly using gloved hands and set the bowl aside for 20 to 30 minutes.

3. Toss the mixture again. Fill the jars one by one with the mixture until you have about ½ inch headspace. Press the mixture firmly into the jars. You may not need the third jar. If you do need it and it's not full, don't worry; the vegetables will be safe as long as they are submerged.

4. With the vegetables packed in tightly, add filtered or spring water to each jar to just cover the vegetables. Poke down the mixture with a wooden chopstick to get out any air bubbles, put on the lids and bands, and screw down until you meet resistance, but not tightly. Set these on the kitchen counter for three days, then tighten the lids and store them in the fridge for up to 6 to 8 weeks.

➤ NOTE: Make half the peppers hot ones like jalapeños, serranos, Thai bird peppers, and habaneros (if you can stand the vicious heat). Use any sweet pepper you like, but for best flavor, make sure they are ripe—red, yellow, or orange, not green. The Anaheim chile, with its mild spiciness, is ideal as a substitute for sweet peppers.

Moroccan Preserved Lemons

These preserved lemons add a tantalizing aroma and a sharply defined flavor to tagines of all kinds, but especially North African lamb stews. Flavorful bits of fermented lemon bring simple vegetable dishes to life, too. Add some to a tapenade, mash some into butter and use it to flavor grilled chicken or fish, add it to chopped raisins used in a couscous. Because it's the peel that you use after the fermentation, make sure your lemons are organic and well washed before processing. Ordinary Eureka lemons are fine, but Meyer lemons are even better, as they have a sweeter, more delicate flavor.

Makes 2 to 3 quarts

12 organic lemons
½ cup sea salt
4 whole cloves
5 black peppercorns
1 bay leaf, crumbled
8 coriander seeds

1. Wash and dry the lemons. Place 1 tablespoon of the salt in the bottom of a sterilized quart canning jar. Mix the cloves, peppercorns, bay leaf, and coriander seeds together in a small bowl.

2. Remove the hard bit at the stem end of the first lemon. Starting at the blossom end, cut through the lemon down toward the stem end until you are an inch from the end. Rotate the lemon 90 degrees and make a second cut, so that you've cut an X to within an inch of the stem end. Pack the interior of the lemon with about 1 tablespoon salt and a pinch of the spices, squeeze the X closed, and put the lemon into the jar.

3. Repeat with more lemons, adding salt and layering in the spices each time, pushing each down into the jar so as to make the juice run. When you're an inch from the top, stop. Put on a sterilized lid and screw it down with the band. Set the jar aside. The next day, push the lemons down again, and repeat on the third day. If on the third day the juice hasn't completely covered the lemons, squeeze out juice from the remaining lemons of your original dozen and add the juice to the jar so the lemons are covered. Don't use commercial bottled lemon juice or water.

4. Set the jar, with the lid just barely closed so any gases can escape, on a warm kitchen counter for 30 days, pushing down the lemons in the jar with a sterilized knife a few times during this period. If you see a cloudy jellylike "mother" form on top, simply remove it. If fuzzy growth appears, that's mold. Start over.

5. Store your lemons in the fridge after the 30-day period. To use, sterilize a fork and take out a lemon, placing it in a bowl. Scrape off the pulp but reserve the salty juice to add back to the lemon jar. Rinse the peel to remove some salt, then cut the peel into thin strips or dice into small cubes.

Fermented Hot Sauce

How hot do you like it? It can be nice and hot using serrano or jalapeño peppers, or down-on-all-fours-pounding-the-floor-and-crying hot if you use habaneros or pequins. You decide. Personally, I like to use habaneros, or their close cousins, Scotch bonnets, because they have a delicious, fruity aroma and—I'm assuming—a similar flavor, although they are so damn hot even a little obliterates your taste buds. So I use only a drop or two of this hot sauce on huevos rancheros, frittatas, in a soup or stew, or to make the devils dance on Asian dishes. I would caution you to use goggles and protective kitchen gloves or even vinyl medical gloves when handling these peppers. A tiny splash in the eye is no pleasant experience. Even touching your eye with a finger that's touched a habanero is pain city.

Makes 1 scant quart

> 3 pounds fresh hot chiles
> 6 cloves garlic, peeled and rough chopped
> 2 tablespoons Sucanat (whole cane sugar)
> 2 teaspoons sea salt
> 1 tablespoon homemade kefir (pages 163–68) or whey mixed into ¼ cup
>   lukewarm water

1. Trim the stems from the chiles, but if using habaneros or Scotch bonnets, leave the green tops.

2. Place all the ingredients in a food processor or blender and process to a smooth paste.

3. Transfer the paste to a quart canning jar, cover the top with a square of paper toweling, and screw it down with a canning band. Allow it to sit at room temperature for 1 week, stirring it once or twice during that time.

4. Place a fine-mesh strainer over a bowl and turn the paste into the strainer. Using a spoon or spatula, press the paste into the strainer, scraping back and forth so the hot sauce runs into the mixing bowl, leaving bits of seeds and pulp behind.

5. Pour the hot sauce into a jar, close with a lid, and store for up to 2 to 3 months in the fridge. Use sparingly.

Horseradish Sauce

Horseradish is easy to grow yourself, and usually available as a root at markets like Whole Foods. When you make this sauce from a fresh root, it will clear your sinuses quickly! It also gets cultured by the bacteria in fresh whey. You'll need to use filtered or spring water. It's great with pot roast or other beef dishes, the way the Germans like it. Mix some with chili sauce to make cocktail sauce for cold Gulf shrimp.

Makes about 1 cup

> **1 cup peeled, shredded fresh horseradish root**
> **1½ teaspoons sea salt**
> **¼ cup fresh whey**

1. Peel and shred the horseradish root and put it in a blender. Add the salt and whey. Blend into a thick puree. Then add water a little at a time, blending until a smooth sauce forms.

2. Transfer the sauce into a small jar with a lid loosely closed. Allow the jar to sit in a kitchen cabinet at room temperature for 1 week. Refrigerate the jar after the fermentation. It will keep for 2 to 3 months.

CHAPTER 13

The Grain and Flour Ferments

Bread Starter

Is it possible to attain terroir in a homemade loaf of bread? Well, every kitchen has its own unique mix of microorganisms floating in the air, so if you capture these yeasts and bacteria by putting a bowl of flour and water out for them, you will soon have a one-of-a-kind fermenting mixture that will leaven your bread. In fact, before people started producing commercial yeasts at the end of the eighteenth and beginning of the nineteenth centuries, bread was made at home (or at community ovens) using the householder's own starter that was sometimes handed down through generations. The Boudin Bakery in San Francisco uses a starter that was first made in 1849 and has been kept alive and refreshed ever since.

Here's how to make a starter that's uniquely yours. You can use all-purpose or whole wheat flour, but I prefer all-purpose flour, which seems to get the starter off to a stronger *start. Yeast and bacteria in the air will find your starter, but if you want real sourdough bread, Cultures for Health (www.culturesforhealth.com) sells sixteen heirloom varieties of sourdough starter.*

When your starter is being colonized by your ambient microbes, its pH will slowly drop and it will become more acidic as naturally occurring enzymes turn starch to sugar, yeast turns sugar to alcohol, and acetobacter turns alcohol to acetic acid. In addition, there will be some lactobacilli in the starter that will add lactic acid to the mix. The acidic environment and the overwhelming presence of yeast cells will almost surely protect the starter from harmful microbes, but my rule of thumb is when in doubt, throw it out. The starter should smell yeasty and clean, with a light note of vinegar, and with no black or gray mold spots.

1 cup unbleached all-purpose or whole wheat flour, preferably organic
1 cup filtered or spring water

1. Mix the flour and water in a gallon-size ceramic bowl. Cover with a moist dish towel and set a plate on top to hold the cloth in place.

2. Let the mixture rest on the kitchen counter for 3 to 6 days, stirring it daily and remoistening the dish towel each day.

3. It's ready when it has risen and deflates when poked, and when a spoonful of it shows little bubbles and smells slightly but cleanly sour. It may fail the first time, but persist with a new batch and soon you will have your starter.

4. Place the starter in a canning jar topped with a piece of paper towel held by a band that would ordinarily hold a metal canning jar lid. This allows the starter to breathe. Place the starter in the fridge.

5. Refresh and feed your starter every 3 or 4 weeks. It will be reduced in size because you will have (it's hoped) made bread with some of it. Simply add a slurry made of flour and water in a 1-to-1 ratio. Stir the slurry into the remains of the starter, let it warm, covered, to room temperature for a few hours, then place it back into the fridge. Refrigeration slows fermentation. The starter should always smell fresh, slightly sour, and yeasty. If it develops an off smell or shows any signs of unwholesomeness like gray liquid or mold, or becomes unpleasantly and very strongly sour, toss it and start over with a fresh batch.

6. When it comes time to make bread, you'll need to get your starter super-charged so it's potently active. A couple of days before your planned day to bake bread, take ¼ cup of starter from your fridge and place it in a large bowl. If you have an accurate scale, weigh the starter and add equal weights of flour and water. If you are using measuring cups, the ¼ cup of starter is given ¼ cup of filtered water and a little less than ½ cup of flour. Mix these ingredients together vigorously so they're thoroughly combined and some air is whipped into the starter. Cover and let stand for a minimum of 4 and a maximum of 12 hours. Then repeat the process, using the same amounts of starter, flour, and water. Mix well, cover, and let stand again for 4 to 12 hours. Feed the starter a third time, and at the end of the third period, the starter should be bubbly and able to double in size in from 4 to 8 hours. Now it's working and ready to adequately proof your bread.

7. On baking day, add 1 cup of the supercharged starter to every 6 cups of flour in the bread recipe. If you've made your own starter, be aware that the yeast float-ing in the air that has made your starter is no match for the active dry yeast you buy at the supermarket. It will take your yeast starter longer to proof (rise) your loaves—from 4 to 24 hours, depending on the temperature and the mix of yeasts in your starter. The longer it takes to rise, the more sour the loaf will be. You'll get the hang of timing once you bake a few loaves.

Making Homemade Bread

Because bread dough has been supercharged with yeast, it is improved nutrition-ally. Yeast is an excellent source of protein and vitamins, especially the B-complex vitamins, as well as minerals and other cofactors required by the body to make amino acids for growth. It is also naturally low in fat and sodium.

You'll need some good equipment if you are going to make fine bread at home. The following equipment can be purchased from the Baker's Catalog. (See contact information on page 280.) The catalog itself is full of excellent bread-baking infor-mation as well as tools.

A baking stone really helps give your bread a professional finish. It's a heavy,

MAKE BREAD STARTER USING ORGANIC FRUIT

Here's a way to make bread starter using the yeast that occurs on certain kinds of fruits. The best sources for wild yeasts are fruits that have a whitish bloom on their skins—grapes, elderberries, blueberries, and huckleberries in particular. It doesn't take much fruit to get a culture going, and sugar in the fruit and yeast on the skins mean that nature has everything already set up for you. Use organic fruit; that way you can be guaranteed that the fruit hasn't been sprayed with toxic fungicides or grown using other agricultural chemicals. If you are harvesting fruit from the wild, choose fruits that grow away from sources of toxins like roadsides, old dump sites, and conventional agricultural fields. Old meadows, woods' edges, and burnt-over forest sites are ideal.

(From top left clockwise): grapes, elderberries, huckleberries, blueberries

In a bowl, mix 1 cup of organic all-purpose or whole wheat flour with 1 cup of water. Remove the stems from the fruit and crush the raw fruit in a separate bowl using a potato masher. Strain the crushed fruit and add ½ cup of the juice to the flour and water mixture and allow it to sit, covered with a cloth or paper towel, on the kitchen counter until the mixture bubbles, smells yeasty, and inflates with carbon dioxide given off by the yeast feasting on the fruit sugar— 2 or 3 days. When the fermentation is starting to slow down, 3 or 4 days into the process, transfer this starter to a clean canning jar covered with a piece of paper towel screwed down with a canning lid band and place it in the fridge. Proceed from there following the instructions in Bread Starter (pages 223–25), beginning at step 5. This starter will initially have a barely noticeable fruit taste but will be visibly more vigorous than starter made from ambient microbes because it's getting a stronger shot of yeast from the fruit.

rectangular, ceramic block that fits in your oven and soaks up intense heat. When you slide your bread dough onto it, the stone gives positive bottom heat to your bread, which helps it rise and finish properly. If you don't have a baking stone, you can line a baking sheet with parchment paper. It's good to have a peel—a large, flat-bladed wooden paddle used for sliding dough into the oven, moving loaves as needed, and removing breads when they're done.

You'll need a cutting tool like a razor blade, X-Acto knife, or lame, the latter being a French tool for slashing the tops of your loaves before putting them in the oven. Unless a knife is literally razor sharp, it tears at the dough and partially collapses it.

You may want to buy some dough-rising baskets made of coiled wooden dowels to give your bread a professional beehive look. And you may want to purchase a couche—an untreated baker's cloth made of flax into whose folds you lay your dough for rising. Finally, do you have one of those cast-iron cornbread molds, the ones with eight depressions that turn out cob-shaped cornbread? They're perfect for getting the right amount of steam into the oven for the first ten minutes of the bake.

This recipe takes several days. Time it so that you'll be around all day on baking day. The previous days take only a few minutes' work.

On the first day, in a large bowl, mix 2 cups of all-purpose organic white flour with 2 cups of lukewarm water, ¼ cup of sourdough starter or ½ teaspoon of dried commercial yeast, and 1 heaping tablespoon of sea salt. The salt retards the development of the yeast. Don't be afraid that you're not adding enough yeast. This recipe takes enough time for the scant amount of yeast added this first day to multiply manyfold. Stirring is done with a wooden spoon just until the ingredients are thoroughly mixed. Cover the bowl with a clean dish towel wet with hot tap water and wrung out, with a plate set on top to hold the towel in place. Set the bowl on a warm kitchen counter away from any drafts.

Couche with baguette loaves

On the second day, stir in ½ cup or slightly more lukewarm water, 1 cup of *white* whole wheat flour,

½ cup of rye flour, and a good handful of rolled organic oats. Wet the dish towel again and cover the bowl with the towel and the plate.

On the third day, it's time to bake. Soon after getting up in the morning, take out your largest bowl and mix 5 cups of all-purpose organic white flour with about 2½ cups of lukewarm water—enough so that the dough holds together. Remove this sticky mass to a floured board and gently knead it for 2 to 3 minutes, then return it to the large bowl and let it rest for 30 to 40 minutes. The exact amount of time isn't critical, but the resting period is responsible for heightened flavor in the final bread.

After the resting period, uncover the bowl containing the poolish, as the mixture that's been sitting for two days is called. Now the mixture is supercharged with active yeast and other microbes ready to go into action. Using a spatula, scrape the poolish into the large bowl with the rested dough, getting as much as possible. You may want to remove any rings from your fingers at this point. Using your hands, squeeze and mix the poolish and the new dough together until they are fairly well incorporated. Flouring the hands is more ritual than practicality, for this mixed dough will be powerfully sticky. Wetting your hands works better. With both hands stuck full of dough, transfer as much of the mixture as you can to a board heavily floured with all-purpose white flour. Scrape the sticky stuff off your hands and fingers and push it into the dough, which will be runny—and may even try to run off the board. Don't let it. Heavily flour the top surface of the dough and push it back into the flour on the board with your hands. Once you're sure that the dough will behave itself and stay put, quickly wash your hands (this will be a relief), dry them, and flour them.

Now fold the dough in half, then gently pull it out to its former size. Give it a half turn, fold, and pull. Use as much flour as you need to keep the surface from being so sticky you can't fold and pull. Knead the dough for 8 to 10 minutes to work up the gluten that causes the bread to be stretchy and chewy. If it begins to stick to the board, use flour to prevent that. A pastry scraper is helpful. At the end of the kneading time, the dough will stay put on your board, although it will still be moist. Go to the sink and thoroughly clean and dry your largest bowl. Wipe the inside of the bowl with a paper towel moistened with a little olive oil. Use both hands to transfer the dough from the board to the bowl. Cover the top with a damp dish towel. Set the bowl aside in a warm (75 to 80°F is ideal), draft-free spot to let the dough proof for at least 4 hours. At the end of that time, it should have doubled or nearly tripled in size.

Gently pull the dough away from the sides of the bowl and deflate it. Cover it again and allow it to rise for another 2 hours, when it will have at least doubled in size, if not tripled.

Gently remove the dough from the bowl and place it on a lightly floured board. If you want 3 smaller loaves, divide it into thirds. If you are making 2 large round loaves, divide the dough into equal halves. If you want baguettes, slice it into 4 equal pieces. Try not to deflate the dough any more than you have to during these procedures.

For 3 loaves, use 3 coiled-dowel baskets well dusted inside with flour. For 2 large round loaves, use 2 round-bottomed baskets about 10 or 11 inches in diameter and about 8 or 9 inches deep. Line them with fine cloth napkins well dusted with flour. If making 4 baguettes, dust the couche with flour to prepare it.

Once the dough is divided, either gently pat it into large rounds and place them with their smoothest side down in the baskets, or gently form them into 4 baguette loaves about 4 inches across and about 12 inches long. These get carried to the couche one by one. Fold the couche into hills and valleys and lay each piece of dough into a valley so that it is given a fold in which to proof. To prevent the couche from flattening out, prop up either end of the flax with a jar or bottle of water—whatever's heavy enough to prevent the couche's folds from flattening out. Then lightly dust the top of the dough, whether basket or couche, with organic yellow cornmeal.

The final rise takes about an hour or two, depending on the strength of your starter. Remove the racks from your oven except for the middle one and the one below it. Place your baking stone on this middle rack and place your cast-iron cornbread mold (or cast-iron skillet if you don't have the mold) on the bottom rack. When the final rise is half over—that is, when there's about a half hour to go—turn on the oven to 500°F. It takes a good half hour for the oven and the baking stone to reach an even 500°F.

After the final rise, have 8 ice cubes ready. Get your razor blade and a glass of water. Quickly open the oven door and sprinkle the stone with a light dusting of cornmeal. Close the door quickly.

For round loaves, have your peel lightly dusted with cornmeal. Place it over the top of the round loaves and invert the baskets. The basket, whether the round dowel type or one lined with a floured cloth, should lift right off. Make your slashes with the lame, dipping it in water between cuts. For round loaves, use 3 slashes that cross in the center. It may take some practice to be able to make quick, sure slashes that don't tear and deflate the dough.

Open the oven door and slide the dough off the peel and onto one side of the stone. Close the door. Slash and insert the second loaf onto the other side of the stone. Toss 1 ice cube into each of the 8 segments of the iron cornbread mold, or all 8 into the iron skillet. Close the door immediately.

For the baguettes, flip a loaf from one end of the couche onto a lightly cornmeal-dusted wooden board about 6 inches wide and at least a foot long, take it to the peel, and slide or flip it on. Repeat until all 4 loaves are on the peel, or place them on the peel and in the oven one by one—whatever seems easiest to you. Using the razor blade, make 4 equally spaced oblique slashes across the top of each long loaf. You can be creative with your slashes if you wish, making one long one, a wavy one, or whatever. Four obliques do the job well and give a pleasing appearance. It helps to dip the blade in water after each slash.

Slide the loaves off the peel onto the stone or baking sheet with a quick back-and-forth shake, quickly add the ice cubes, and close the door. The reason for the ice cubes is the same reason professional ovens have steam injectors. The ice cubes turn to steam and evaporate over the first 10 minutes or so of the bake and produce a beautiful, crisp crust.

The bread should bake at 500°F for 15 minutes. Then turn the oven temperature down to 350°F for another 30 minutes. At that time, your bread should be done. Slide the peel under each loaf and remove it from the oven. Turn it upside down and thump the bottom. It should sound hollow. Set the breads right side up to cool on racks. Round wire-mesh pizza baking screens are ideal, as they let cool air surround the loaves and let them cool evenly. Set the pizza screens atop bowls so air can circulate freely.

Allow the loaves to cool completely before placing them in plastic bags to freeze. When you want to use them, thaw them out at room temperature in their plastic bags.

You may find that you have room for only 1 round or 2 long loaves on your stone. If so, do consecutive bakes. Bake 1 round loaf or 2 narrow loaves as above, allowing the remaining dough to rest in its basket or in the couche. When the first bake is done, turn the oven back up to 500°F and wait 15 minutes before beginning the second bake, which is done just like the first.

You'll find these loaves have a beautiful crust and an internal structure full of holes, with a slightly translucent, stretchy, deliciously chewy texture. The whole wheat, rye, and oats give real flavor to plain white flour. You'll also notice with pleasure that this bread will last for days and days without getting stale.

Many variations are possible. Using more whole wheat will give a denser and more nutritious loaf. More rye flour gives you a heavier rye bread (and stickier dough to work with). Black walnut meats, if freshly shelled and full of aromatics, are wonderful in bread. Hickory nutmeats are even better. Oil-cured olives, pitted and roughly chopped, may be added for little flavor bursts. Garlic cloves can be roughly chopped and added. Fennel seeds, caraway seeds for the rye, pumpkin seeds, sunflower seeds, and other seeds will give a chewier texture and varying flavor layers. But before you start experimenting with additions, perfect your basic bread as given in the recipe above.

Easy Homemade Bread

This recipe was devised by Jim Lahey at Sullivan St Bakery in Manhattan. It produces great bread for very little work. You will need a heavy cast-iron Dutch oven with a cast-iron lid to make it, though. Don't forget to supercharge your starter for bread baking starting three days before baking day (see instructions on pages 225 and 228). Make this dough before going to bed and it will be ready for subsequent steps and baking when you get home from work the next day.

Makes one 24-ounce loaf

3 cups all-purpose flour
¼ teaspoon active dry yeast or ½ cup of your own starter (see pages 223–25)
1½ teaspoons sea salt
Cornmeal as needed

1. In a large bowl, mix the flour, yeast or starter, and salt together. Add 1⅝ cups of water, and if you are using the starter that you've supercharged for bread baking, you can omit the yeast or use just a sprinkle. Stir until it makes a shaggy, sticky dough. Cover the bowl with plastic wrap and set aside on the kitchen counter at about 70°F for about 18 hours. The dough will be ready when it's dotted with bubbles.

2. Lightly flour a board or work surface and turn the dough out onto it. Sprinkle the dough lightly with flour, then fold it over on itself once or twice, leaving it seam side down. Cover the dough loosely with plastic wrap and let it rest for 15 minutes.

3. Place a clean cotton or linen dish towel on the work surface and put a generous amount of cornmeal on it where the dough will sit and rise. Flour your hands, or wet them, and quickly shape the dough into a ball and place it seam side down on the cornmeal. Dust the top of the dough with a little more cornmeal and cover it with another clean dish towel. Let it rise until doubled in size, about 2 hours.

4. Make sure the oven rack is placed so there's enough room to accommodate the Dutch oven with its lid on. Preheat the oven to 450°F. Place the empty covered Dutch oven on the rack while the oven heats.

5. When the dough has risen sufficiently and the pot is thoroughly hot in the oven, pull out the rack and remove the pot lid. Slide your hand under the dish towel and turn the dough upside down into your other hand, quickly brushing off some of the loose cornmeal, then place the dough in the pot seam side up. Immediately cover it with the lid and slide the pot back into the oven on its rack, closing the oven door.

6. Bake for 30 minutes, then remove the lid and bake for another 15 to 30 minutes, until the loaf is browned to your liking. Using two spatulas, lift the finished bread from the pot and set it on a rack to cool.

For Rye Bread

Substitute ½ cup of rye flour for ½ cup of the all-purpose flour.

Rustic Whole Wheat Bread

Some people say that whole wheat bread won't yield that chewy, rustic look because the bran edges slice the gluten into small fragments and you lose that stretchy texture. This recipe, inspired by Amy Scherber's golden loaves at Amy's Bread in New York City, turns that notion on its head and adds more bran to the dough. It's a slack dough and wonderfully tasty. Since it takes about five or more hours with several rises, this bread is best made when you plan to be home for most of the day.

Makes 2 loaves

> **1 teaspoon dry yeast or ½ cup of your own starter (see pages 223–25)**
> **2½ cups whole wheat flour**
> **1½ cups unbleached all-purpose flour**
> **½ cup wheat bran**
> **1 tablespoon sea salt**
> **Olive oil or nonstick vegetable spray**
> **Cornmeal as needed**

1. Place the yeast or starter in a small bowl and add ½ cup lukewarm water. Stir until dissolved. If using starter, make sure it's been supercharged (see directions on pages 225 and 228) for bread baking.

2. In a medium bowl, combine the whole wheat flour, all-purpose flour, bran, and salt. Add the starter or dissolved yeast and 2 cups of cool filtered water or spring water. Use a flexible plastic spatula to mix these ingredients into a homogenous loose mass. Wet your hands and knead the mass in the bowl for about 2 minutes, adding another tablespoon or two of cool water as needed. The dough will be powerfully sticky so you might want to remove your rings before kneading.

3. Coat the inside of a large bowl with olive oil or nonstick vegetable spray and set the bowl aside. Very lightly flour a board and turn the dough onto it. Knead for 3 to 4 minutes, until the dough is smooth, using as little flour as possible. A pastry scraper helps get all the sticky dough off the board when you're turning it, but you can use a regular metal spatula, too. Place the kneaded dough back into the medium bowl and cover loosely with oiled or sprayed plastic wrap. Let the dough rest for 20 minutes.

4. Return the dough to the lightly floured board and knead for 6 to 7 minutes. The dough should be silky soft and loose. Placed the dough into the large oiled bowl, cover the top loosely with oiled plastic wrap, and set aside in the warm kitchen for 1 hour.

5. At the end of the hour, punch down the dough and bring the ends together. Turn it upside down so the seam is down and let rise another hour, covered loosely with oiled plastic wrap.

6. At the end of this second hour, punch it down again and repeat the folding and turning. Cover with plastic wrap as before. Let it rise for about 90 minutes, then poke the dough gently with a finger. It's fully risen when the indentation doesn't spring back.

7. If you have a baking stone, place it on a middle rack in the oven. Place a small cast-iron skillet or cornbread mold on the lowest rack. No need to heat just yet. Sprinkle cornmeal onto a large baker's peel, or if you don't have one, line a baking sheet with parchment paper. Divide the dough into 2 equal pieces. Shape them with your hands into slightly flattened balls and set them on the peel or baking sheet at least 4 or 5 inches apart. Cover with oiled plastic wrap, as before, and allow them to rise for another 90 minutes. At the 60-minute mark during this last rise, preheat the oven to 450°F.

8. Using a razor blade, very sharp knife, or lame, quickly slice a shallow 3-slice star pattern in the top of each loaf.

9. Have 8 ice cubes ready. Open the oven door and place the ice cubes in the skillet or a cube in each of the 8 sections of the cornbread mold. Slide the loaves onto the stone or place the baking sheet with the unbaked loaves on the rack. Quickly close the door. Bake for 15 minutes, then reduce the oven temperature to 375°F. Bake another 15 to 18 minutes, until golden brown and the loaves sound hollow when tapped on the bottom. Loaves baked on the baking sheet will take a few minutes longer than loaves baked on the hot baking stone. Cool on a wire rack.

A Kefir Starter for a Sour Loaf

Once you have your kefir grains, you have a daily supply of kefir, and when bread-making time comes around, you can use kefir as a starter. It's extra nutritious and gives the bread a new layer of tempting aroma—as if bread needed any. Substitute 1 cup of rolled oats and 1 cup of rye flour for 2 cups of the all-purpose flour for richer-tasting loaves. You can try this starter exactly like a starter made from wild microbes from the air, or as you would use yeast from a packet. This recipe yields enough starter for 3 or 4 loaves made from 8 cups of all-purpose flour. Prepare this starter in the evening before the day when you'll be making bread.

Makes 3 or 4 loaves

Starter

    2 cups organic unbleached all-purpose flour
    1 cup kefir (pages 163–68)

Dough

    8 cups organic unbleached all-purpose flour
    1 tablespoon sea salt
    1 packet bread yeast
    1 teaspoon honey

2 tablespoons extra-virgin olive oil

1 cup warm spring or filtered water

Heavy cream for brushing tops, if desired

1. Make the starter: Place the flour in a large bowl and make a well in the center. Pour the kefir into the well.

2. Bring flour from around the well into the kefir and mix until it's thoroughly incorporated. Then turn it onto a lightly floured board and knead until it's smooth and elastic, about 5 minutes.

3. Place it back into the bowl and cover the bowl with a damp clean dishcloth or plastic wrap.

4. The next day, knock it down and set it aside to use in your favorite bread recipe, or continue with the recipe to make your bread.

5. Make the bread: Place the starter in a large bowl and add the flour, salt, yeast, honey, and oil. Add the water a little at a time until you have a smooth dough for kneading.

6. Knead the dough for 8 minutes, or until it's smooth and elastic, then cover the bowl with a damp cloth and allow it to rise in a warm place until it's doubled in size; the time will depend on the temperature.

7. Punch down the dough, replace the cloth, and let it rise until doubled again.

8. Grease 3 or 4 loaf pans and divide the dough into 3 or 4 parts. Spray pieces of plastic wrap with nonstick cooking spray and let them settle over the pans, oiled side down against the top of the dough.

9. Preheat the oven to 425°F.

10. When the dough is nicely raised, make a slice or two diagonally across the top of the dough with a scalpel, razor blade, lame, or extremely sharp knife with quick gentle strokes so you don't collapse the dough. Brush the tops with cream if you wish, but not in the slashes, and bake for 35 minutes for 4 loaves or 40 to 45 minutes for 3 larger ones, or until a loaf sounds hollow when pulled from a pan and tapped on the bottom. Cool on a wire rack.

Pizza Dough

This dough requires a starter that is going at peak power. See the starter recipe on pages 223–25 to get your starter up to speed. While the microbes are killed off during the bake, you still get the benefit of their metabolic products along with your gooey pizza.

Makes 1 large pizza dough

> **1½ cups sourdough starter (pages 223–25)**
> **5 tablespoons extra-virgin olive oil**
> **½ teaspoon sea salt**
> **About 1½ cups all-purpose flour**

1. In a bowl, mix the sourdough starter, 1 tablespoon of the olive oil, the salt, and 1 cup of the flour. Work more flour into the dough a little at a time until a smooth dough is formed. How much flour is used depends on how wet your starter is.

2. When you achieve a smooth, pliable dough, set the bowl aside for 30 minutes.

3. Place a pizza screen or baking sheet in the oven and preheat to 500°F. On a lightly floured board, roll the dough into a circle, using only the minimum flour to prevent the dough from sticking.

4. Lay the rolling pin near one edge of the dough and flip the edge toward you, over the pin. Roll up the circle of dough onto the pin. Open the oven door and unroll the dough onto the pizza screen or baking sheet. Close the door and bake for 7 minutes.

5. Take the crust on its screen or baking sheet from the oven and brush its top surface all over with the remaining olive oil. This prevents the toppings from dribbling juices into the crust and making it soggy. Add your toppings and return the pizza to the oven on its screen or sheet and bake until the cheese melts, the edges of the dough are nicely browned, and the toppings are cooked.

Kishk el Khameer

Kishk el khameer is a lacto-fermented form of bulgur, although sometimes small North African couscous is used. As the bacilli work on the cracked wheat, and as it's kneaded frequently over many days, it takes on a cheesy aroma, a tangy flavor, and a texture like goat cheese. It originated in the Middle East probably millennia ago, although written records of it date to the thirteenth century. Poor farmers made kishk with water, but wealthier people kneaded the bulgur with yogurt, and today some people recommend making it with kefir or buttermilk.

Kishk el khameer is a traditional food in south Lebanon, and it has come to the attention of Slow Food and been added to that organization's presidium list as an endangered food that's being commercially produced. Vegans have discovered it, when made with water, as a non-dairy substitute for cheese.

It is made into two forms. The fresh form, called green kishk, can be used as soon as it ripens, or it can be rolled into little balls the size of large marbles and stored in olive oil. A second form is made by drying it in an oven with a pilot light, in a food dehydrator, or on a roof where the climate is sunny and dry. Think of the climate of Lebanon or Southern California. In humid regions, mold will grow on it, ruining the batch. When dry, it's crushed into a powder that's used as a flavor and thickening agent for soups and stews.

I've made it with kefir and found that the balls stored in oil make a nice addition to tabouleh. They have a very distinctive and pleasant flavor. In the village of Majdelyoun in southern Lebanon, where small amounts are made commercially, it is flavored with herbs and spices. The following recipe omits the flavorings so that you can taste it in its basic form, but feel free to experiment, especially with the herbs of the dry Mediterranean countries: thyme, rosemary, oregano, and sage.

Makes about 18 balls

> 1 cup filtered water, yogurt (pages 174–77), or kefir (pages 163–68)
> ½ cup bulgur or couscous
> 1 tablespoon sea salt

1. Place the liquid in a ceramic bowl and mix in the bulgur. Cover with a clean dish towel.

2. Stir the mixture once a day for 8 days, recovering it with the towel each time.

3. At the end of 8 days, add the salt and knead it into the kishk.

4. Place the kishk into a sealed container kept at room temperature and take out the dough to knead it for a few minutes every other day for 30 days. At the end of this period, it should be ripe.

5. Use it fresh or roll it into balls the size of marbles, place them in a sealed container, and cover them with olive oil for storage. Store in the fridge to stop further ripening.

Injera

In Ethiopia and Eritrea, in East Africa, this spongy, sour flatbread is used to scoop up saucy meat and vegetable dishes, much as naan is used with Indian food. Injera is also used to line the plates on which the stews are served, and its bubbly texture soaks up and holds the juices. When the stews are all picked up by torn pieces of injera and the plate liner is torn apart and eaten, the meal is over.

Injera is made with teff, a tiny, round grain that flourishes in the highlands of Ethiopia. While teff is very nutritious, it contains practically no gluten. This makes teff ill-suited for making raised bread, but perfect for folks who are gluten intolerant, so if you are gluten intolerant, use 100 percent teff flour in this recipe. Teff flour can be hard to find in some regions, but check a well-stocked health food store, Whole Foods, or order online at www .myspicesage.com. It is also fermented, but without the leavening properties that wheat gluten imparts. Fermentation with yeast gives it an airy, bubbly texture, and also a slightly sour taste that I would guess is caused by the presence of lactobacilli along with yeast.

The injera served in many East African restaurants here in the States often includes both teff and wheat flour. Most injera made in Ethiopia and Eritrea, on the other hand, is made solely with teff. I know an Eritrean family who lives in San Pablo in the East Bay re-

gion of San Francisco, and the very lovely woman who makes injera at home complains that whether she uses all teff or teff and wheat flour, it just doesn't taste like it does in Eritrea.

Well, sourdough bread made in Chicago won't taste like the San Francisco kind, either. The flavors of injera and sourdough and many other breads are dictated in large part by the indigenous microbes floating in the air as well as the ingredients.

Like making pancakes, the first one may not be perfect, but you'll soon have the hang of it. Serve with browned lamb or beef chunks, or boned chicken chunks, in a curry or tikka masala sauce with Sriracha sauce to spice it up.

Makes 4 to 6 injera

½ cup teff flour
½ cup all-purpose flour
1 cup water
Pinch of salt
Peanut or vegetable oil

1. Place the teff flour in a large bowl. Sift on the all-purpose flour and stir to mix. Add the water, stirring constantly to avoid lumps, until a smooth mixture is formed.

2. Set the bowl aside on a warm kitchen counter covered with a clean dish towel for 2 to 3 days, until it's bubbly and fermenting by the wild yeast that's colonized it, turning it slightly tangy. If it fails to bubble after 3 days, stir in 1 teaspoon of commercial yeast.

3. After fermenting, add the salt and mix to incorporate.

4. Find your largest, smoothest-surfaced pan or skillet and lightly oil it. Heat it over medium heat until a water drop dances on the surface. Spoon about 4 to 5 tablespoons of batter into the center of the pan, and tilt it gently in all directions so the batter thins out.

5. Watch for bubbles to form on the surface of the injera. When they've broken and the surface of the flatbread is just about dry, remove from the heat and lift off the flatbread with a spatula. Injera should be thicker than a crepe but thinner than a pancake. Repeat until the batter is gone. Serve hot.

CHAPTER 14

The Bean and Seed Ferments

Homemade Miso

Miso is the wonderfully nutritious and tasty fermented soybean and rice paste that the Japanese use to make their staple miso soup. You can make it yourself at home, but you need the special fungus, called koji in Japan, whose scientific name is Aspergillus oryzae. If that name sets off alarm bells, it's because Aspergillus oryzae's cousin, Aspergillus flavus, is the source of the potent carcinogen aflatoxin. However, the koji mold does not produce aflatoxin. It's been used for a thousand years in Japan to make not only miso, but also sake, amazake, bean paste, and other foods, all without recorded incidence of illness other than in the infrequent person who shows an allergic reaction to the soybeans used with it. For those people, it can be made by substituting lima beans or garbanzo beans. The

following recipe also works fine with either canned or dried garbanzos. Canned garbanzos need no cooking, but they do need a good rinse to wash the salty packing water off them. If using dried garbanzos—also called chickpeas—figure that a cup of dried garbanzos is equivalent to about 2 or 2½ cups of cooked beans. Put the beans in filtered water and soak overnight. Pour off the water, add fresh filtered water to cover, and set to a medium boil for 3 hours. You don't need to add salt to the water, as you'll be using lots of salt when you make miso. The garbanzos will be done when a bean squishes fully when pressed between thumb and forefinger.

While you can use your homemade miso to make miso soup, you may find that it has many other uses in your kitchen. You'll find some ideas for using it after the recipe for homemade miso.

Look for koji rice (dried rice colonized by Aspergillus oryzae) in Asian and Japanese markets, or order it by mail from an online source, such as South River Miso Company: www.southrivermiso.com. It's possible to get koji fungus spores and inoculate the rice yourself, but it's a lot easier to buy the koji rice that's already colonized.

Look for organic soybeans, which are not allowed to be genetically modified. It's a shame that the FDA has not promulgated a rule that food labels must state if they are genetically modified or contain genetically modified organisms (GMOs). Monsanto has patented a GMO soybean that can withstand applications of Roundup (glyphosate) herbicide, and currently about 90 percent of American soybeans have been so modified. I personally don't want to put GMO foods into my system, and so the only way to guarantee that a food like soybeans is non-GMO is to buy organic.

For the fermentation, you'll need a small glazed ceramic crock with no cracks. Mine is a 3-gallon crock about 2 feet tall and 18 inches in diameter. Wash all your utensils well and pour boiling water over them to prevent unwanted microbes or spores. Wash your hands well before handling the food. Read all these instructions before starting.

Your homemade miso will have great power to warm you when made into soup. Its restorative powers make it a favorite for midwives to give to laboring mothers. Because you chose organic soybeans, it has no preservatives, and it hasn't been pasteurized. You'll find it has a sharper, cleaner umami flavor than most mass-produced kinds. This makes it excel-

lent for any broth-based soup, as a glaze for fish or pork, and for some of the recipes that follow.

Makes about 4 quarts

> **1 pound whole dry organic soybeans**
> **3 cups spring or filtered water**
> **5½ ounces sea salt, about ¾ cup**
> **11 ounces dried koji rice, about a quart**

1. Soak the soybeans for 3 hours in the water, until they about double in size.

2. Add the beans and soaking water to a pressure cooker and cook for 40 minutes, using your heaviest weight on the top of the cooker. If you don't have a pressure cooker, boil the beans for 4 hours, or until they are very soft, adding more water to maintain the original level. Allow the pressure cooker to cool, then open the pressure cooker and transfer the beans to a colander set over a bowl to catch the cooking water. Set the cooking liquid aside and put the beans in a bowl or stainless-steel pot for mashing.

3. While the beans are still hot, use a potato masher or—even better—a wooden mallet plunged headfirst into the beans, to crush them until just a few beans are still whole. Allow this mashed mixture to cool to lukewarm (about 95°F, or close to body temperature).

4. Reserve 2 teaspoons of the salt and dissolve the rest in about ¾ cup of the cooking water. Add the salty water to the soybeans, mixing it in thoroughly.

5. Using your hands, pour the koji rice into the bean mixture and work it through the paste until it's thoroughly incorporated.

6. Wet the inside of the ceramic crock and rub it with 1 teaspoon of the salt. Add the prepared miso mixture. Level the surface and sprinkle it all over with the second reserved teaspoon of salt to prevent the growth of unwanted organisms. Cover the surface with a piece of wax paper cut roughly to size. Set a plate that just fits into the container on the wax paper. Weight it with about 10 pounds of well-washed bricks placed in a plastic bag. I found that a 1-gallon plastic freezer bag can be zip-locked

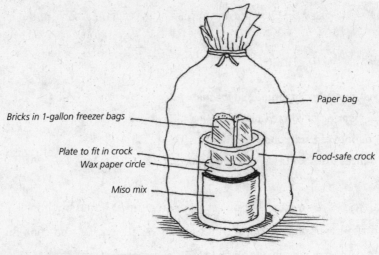

Setting up the miso crock

closed over one brick, so I put bricks into two of these bags and set them on the plate. This makes a small batch, and small batches are easiest to process. If you want to make a larger batch and fill your crock up to three-quarters full, repeat steps 1 through 5 as needed, following the remaining steps over the next day or two until the crock is at least three-quarters, or a little more, full. I have made a nearly full crock at one time by quadrupling the above recipe, but it's much more difficult than making smaller batches. If you do quadruple the recipe, be prepared to spend several hours mashing and mixing.

7. When you remove the weights, plate, and wax paper to add more of the miso mixture, rub the inside exposed surface of the crock with a little salt and sprinkle just a little more salt on the new surface, then replace the wax paper, plate, and weights. Place the crock in a paper bag and close the bag with a length of string. If the crock and bricks are too tall for the paper bag to close, slip a tall plastic kitchen garbage bag over the top and tie it tightly to the paper bag with string. The bag with the crock should be placed in a cool room or, even better, in a dark closet. After a month or two, open the bag, take off the weights, plate, and wax paper and check to see if liquid tamari has formed on the surface. If it hasn't formed, increase the weight. If it has, mix it down thoroughly into the miso, wipe the inside exposed walls of the crock with sterile cloth, and rub just a little salt onto the walls and sprinkle a scant ½ teaspoon on the new surface. Replace the wax paper with a fresh piece of wax paper, the plate, weights, and bag, and set it back in the closet for another 9 months, until it is a year old. (For red miso, store for up to 3 years; the longer it ferments, the darker the color.) Pour off the tamari and store it in the fridge. Taste the miso to make sure it's to your satisfaction. It should smell clean and have good color. After it's finished, store the miso in 1-pound tubs in the fridge or store some for future use in the freezer. It will keep for 2 years or more in the fridge or indefinitely in the freezer.

Miso Salad Dressing or Asian Marinade for Chicken

Makes about ½ cup, enough for 3 to 4 pieces of chicken

> *1 tablespoon miso (pages 241–45)*
> *2 tablespoons rice vinegar*
> *2 tablespoons tamari*
> *2 tablespoons toasted sesame oil*
> *½ teaspoon minced ginger*
> *½ teaspoon minced garlic*

Place all the ingredients in a bowl and mix thoroughly. If you marinate chicken in the sauce, discard the sauce after marination (don't use it for salad dressing). Marinate the chicken in the fridge for 4 to 12 hours before cooking.

Marinade for Grilled Beef

Makes about 1½ cups

> *½ cup miso (pages 241–45)*
> *½ cup plain yogurt (pages 174–77)*
> *3 tablespoons tamari*
> *1 tablespoon minced garlic*
> *1 pound beef steak*

Mix the miso, yogurt, tamari, and garlic thoroughly in a bowl, then pour the mixture into a zip-top plastic freezer bag. Add the steak. Close the bag to remove air and marinate for 6 to 8 hours in the fridge (place the bag in the fridge before going to work and it will be ready by dinnertime); remove the beef from the marinade and discard the marinade before cooking. Allow the steak to come to room temperature before grilling.

Miso-Seasoned Butter

This butter can be used to enliven mac and cheese, to smear on Italian bread, to melt over hot pasta, to enhance hot vegetables, and to melt into hot baked potatoes. Use your kitchen creativity to find other uses.

Makes about ¾ cup

¼ cup miso (pages 241–45)
½ cup (1 stick) unsalted butter, at room temperature
2 cloves garlic, mashed through a garlic press

In a bowl, stir all the ingredients together until thoroughly mixed. Place the bowl in a plastic bag, twist the tie shut, and store in the fridge for up to 1 week.

Homemade Amazake

Amazake is a traditional Japanese fermented food, simple to make, and when made by you at home, free of any chemicals, preservatives, or other unwanted substances. It also uses koji rice to inoculate brown rice with Aspergillus oryzae. The fungus colonizes the rice grains, turning the starches to sweet sugars, and so amazake is used as a hot drink and also as a dessert, snack, sweetener, infant food, and salad dressing, and with pureed fruits as a smoothie. The drink is still served at rituals and traditional celebrations in Japan. It

helps to have a food thermometer when making amazake, as temperatures must be well controlled. It's recommended you make this when you'll be at home all day.

Makes 8 to 9 cups of amazake base

3 cups short-grain brown rice
1½ cups dry brown koji rice

1. Place the brown rice (not the koji rice), in a large saucepan with 6 cups of plain water (no salt) and bring to a full boil over high heat. Reduce the heat to low, cover, and cook for 60 minutes. Stir it up thoroughly and transfer it to a Pyrex or ceramic bowl. Allow it to cool to warm (110°F).

2. Add the dry brown koji rice to the cooked rice and stir it in completely. The bowl should be slightly less than full. The mixture will be thick but will become thinner as it ferments. Cover the bowl with a plate to conserve the heat.

3. Place the bowl in a warm place for 5 to 8 hours. The temperature should remain at about 110°F. You can set it in a warm oven, or place it on the stove over a large pan of water kept on low heat—like a double boiler. But monitor the temperature of the koji frequently. It should not be higher than about 115°F.

4. Stir the rice mixture with a wooden spoon several times during the fermentation, checking the temperature each time. Taste it after 5 hours. If you like the sweetness, you can stop the fermentation there. If you want it a little sweeter, let it continue for the full 8 hours. It should be sweet and the rice grains soft.

5. Stop the fermentation by simmering the amazake over a flame or electric burner for 15 minutes. Simmer—don't scorch. This is now your amazake base, which you can store in a closed jar in the fridge.

To make an amazake drink

Mix 1 part amazake base with 1½ parts water in a saucepan. Bring just to a boil. Pour into heated cups, and top each cup with a pinch of grated fresh ginger.

Other uses for amazake

As a sweetener, substitute 3½ tablespoons amazake for 1 tablespoon honey or 2 table-spoons sugar. Use in breads, cakes, pancakes, waffles, or muffins. It will augment the leavening, add moistness, and sweeten the bread or pastries.

Tempeh

On the island of Java in Indonesia, two indigenous species of mold—Rhizopus oryzae and Rhizopus oligosporus—are typically found on hibiscus leaves. The Javanese have been using an infusion of hibiscus leaves to ferment soybeans for hundreds of years, producing a food they call tempeh. Unlike most fermentations, these molds, rather than bacteria or yeasts, do the work. The result is a very nutritious food. First of all, tempeh is a complete protein with all the amino acids, thus can substitute for meat in the diet. It has many health-promoting compounds like isoflavones and saponins. The molds also produce natural antibiotic agents. The fiber in the soybeans is left intact by the fermentation and there are benefits for the digestive system from enzymes produced during the ferment.

Making tempeh at home isn't difficult, but you do need to have the correct starter of one or both of these two healthful molds. Tempeh starter is available through www.cultures forhealth.com and other online sources.

Tempeh's nutty, mushroomy, umami flavor is good as is, but many cooks like to mari-nate it. Slice it thinly, and deep-fry until crisp. That's the way they like it in Indonesia. But you can also stir-fry, bake, or grill it. It's easy to grate or shred, and when put through a food processor, it acquires the texture of hamburger. Some cooks use it as a hamburger helper, mixing it half and half with ground beef or ground chicken thigh meat. It's a very versatile food.

Makes 2 cakes, each not quite 1 pound

> **1 pound dry organic soybeans**
> **5 tablespoons white vinegar**
> **1 teaspoon tempeh starter**

1. Soak the soybeans in a gallon of water overnight. Unless you bought hulled soybeans, you'll have to remove the hulls. After the soak, squish the beans with your fingers, loosening the hulls so they float free. Make sure the beans split in half as you press them. Pour the water and hulls into a strainer, add fresh water to the pot, and repeat until the hulls are removed. You don't have to get every last one, but try.

2. Drain the beans and place them in a cooking pot, covering them with fresh water. Add the vinegar and turn up the heat. Boil the beans for 30 minutes, adding a little more water if some beans become uncovered.

3. Pour off the water and return the beans to medium heat, stirring to let excess water boil off, until the beans are moist but free of liquid, about 5 minutes. Don't let them scorch.

4. Set the beans aside to cool until they are just lukewarm. Sprinkle the beans with the tempeh starter and stir thoroughly. It's important to work the starter through all the beans so the mold colonizes every part and reduces the opportunity for rogue molds to take hold. Thorough mixing also hastens the fermentation.

5. Take 2 quart-size zip-top freezer bags and puncture them with a clean ice pick so the holes are half an inch apart in all directions. The mold needs air to breathe.

6. Divide the inoculated soybeans in half and put each half into a separate bag. The beans should be about 1 inch to 1½ inches thick in their bags.

7. If you have a gas oven with a pilot light, place the beans in the oven on a baking sheet for 36 to 48 hours. They need warmth, about 85°F, to ferment properly. If you don't have a warm gas oven, place them on a baking sheet set on a heating pad at the lowest setting and cover loosely with a dish towel. Use a thermometer to check the temperature, making sure it stays about 85°F. At the end of the fermentation, the beans should be covered with the mold's white mycelia—slender white threads interwoven on the surface of the tempeh. Remove the cakes from the bags and store them wrapped in wax paper in the cheese drawer of your fridge.

A Marinade for Tempeh

Cut your homemade tempeh into small cubes or slices and soak them for 2 to 3 hours in this marinade. Then add to stir-fries or mix with steamed vegetables.

Makes about ⅔ cup marinade, enough for 1-pound cake of tempeh

2 tablespoons tamari
Juice of 1 lemon
1 teaspoon crushed or ground coriander
1 clove garlic, mashed through a garlic press
Salt and freshly ground black pepper to taste
½ cup filtered or spring water

In a small bowl, mix all the ingredients well and pour into a zip-top freezer bag. Add the tempeh and zip the top closed.

OPTIONS: Here's a chance to get creative. Sake, barbecue sauce, mustard, lemongrass, ginger, tamarind sauce, rice vinegar—all these and more can be used to flavor tempeh. Be selective and restrained.

Natto

Natto—a Japanese fermented soybean—is an acquired taste, to say the least. Jennifer Harris introduced me to it by saying, "You want to try something really nasty? It's sticky and stringy and smells weird, but I love it." I reasoned that some really stinky cheeses like Esrom, Limburger, and Roquefort are among my favorites, so I gingerly tried a few of the fermented beans. The flavor wasn't as strong as the odor, but the texture of the beans' sticky, stringy surface was off-putting. Yet the more I learn about natto, and how it's been favored in Japan for at least two thousand years, the more intrigued I am.

The microbe that turns soybeans into natto is Bacillus subtilis—a soil bacteria by nature that can persist through extremes of environmental hardships. It withstands the heat of cooking, drought, high salinity, extreme pHs, radiation, and solvents. Studies over the past century have shown that it can strongly stimulate the immune system and encourage the production of white blood cells, and it figures in the production of substances toxic to cancer cells. From World War II on, the bacteria was marketed in America and Europe as a remedy for intestinal and urinary tract infections, until cheap antibiotics came into use. The difference was that B. subtilis didn't wipe out the gut's ecosystem the way antibiotics do. But most people weren't thinking about the state of their intestinal flora in the second half of the twentieth century. Today we are, and hence the resurgence of interest in natto, not only as a remedy for illness, but as a very healthful food. Natto contains high quantities of vitamin K and vitamin B_{12}. Vitamin K is much more than an aid to blood coagulation. It has many important effects in the body, including strengthening bone structure. There are several variant forms of vitamin K, but the most healthful is vitamin K_2—menaquinone—and that's the variant found in natto—up to ten times the concentration found in spinach and other leafy green vegetables. Vitamin B_{12}, found in meat, is hard to come by in vegetarian and vegan diets, so natto may be an important source of this essential vitamin. B_{12} is involved in the normal functioning of the brain and nervous system and in the formation of blood. It's also involved in the metabolism of every cell of the body, especially affecting DNA synthesis and regulation but also essential fatty acid synthesis and energy production.

It's possible to make natto at home. I'd advise looking for a source of commercially made natto in a health food store or Japanese market and try it to see if you're intrigued enough to make a batch. If you are, you can use a few tablespoons of the commercial natto to start your own ferment, find spores at alternative pharmacies, or order spores of the bacillus from www.culturesforhealth.com.

As with winemaking, cleanliness is important so that your soybeans are colonized by B. subtilis and not other, opportunistic, bacteria. Sterilize your utensils for ten minutes in a boiling water bath. You'll need a food thermometer (sterilize it by wiping it down with alcohol rather than boiling it, then wiping off the alcohol with a clean paper towel), a stainless-steel or enameled cooking pot, a Pyrex or ceramic casserole dish, a colander, and some kind of enclosed or insulated space that you can keep warm, such as an oven or a *picnic cooler. A lightbulb will provide enough warmth, but make sure the space is roomy enough that the hot bulb doesn't touch the sides. Note that the process takes several days.*

Makes about 2 quarts of natto.

1 pound small organic dried soybeans
2 to 3 tablespoons sterilized (boiled and cooled) water
Pinch B. subtilis powder or 2 tablespoons commercial natto
¼ teaspoon sea salt or pickling salt
2 teaspoons Sucanat (whole cane sugar)

1. Wash the soybeans well, then soak them for 24 hours in four times their volume of water. Drain them and rinse them well. Cook in a pressure cooker using your heaviest weight for 30 to 40 minutes or boil them gently for 4 to 6 hours. The soybeans are done when they can be easily and thoroughly squished between thumb and forefinger.

2 Mix the sterilized and cooled water with the natto powder or commercial natto, salt, and Sucanat. Stir together to incorporate.

3. Drain the soybeans and spread them evenly in the casserole dish, then sprinkle the beans evenly with the natto mixture.

4. Cover the casserole dish with aluminum foil. Poke three or four holes in the foil with a knife to allow air in and out. If no air can exchange, the natto will become very strong smelling. Allowing too much air will allow the beans to dry out and the natto bacilli will stop working. You have to maintain humidity in the casserole dish.

5. Place a lamp without its shade with a regular 100-watt incandescent lightbulb in the oven. Don't turn on the oven—the lightbulb will provide heat. Set an oven thermometer and a moist kitchen towel on a rack away from the lightbulb. Then place your casserole with the natto in the oven. Maintain the temperature at 100°F and maintain humidity by an occasional spray of distilled water—just a quick spritz away from the lamp onto the moist towel. Maintain the heat and humidity for 24 hours.

6. After 24 hours, remove the foil and stir the beans. If the bacilli did its job, the beans will become a bit sticky and stringy. It should not smell or taste sour. Place the beans in a covered container in the fridge for a few days. If all appears well, try some. Use a few for the next batch. For longer storage, natto fermented soybeans freeze well.

7. If the beans have soured or you see mold growing, discard the batch and start over. Like any fermented product, some batches will be better than others. Be scrupulously clean, and chances are, your next batch will be fine.

CHAPTER 15

Fermented Beverages

Water Kefir

Milk kefir grains look like cauliflower and metabolize the lactose in milk, but there is another, closely related, symbiotic combination of bacteria and yeast that works in sweetened water. Folks who can't tolerate any kind of milk products, vegans, or others who want to avoid milk can get the benefits of kefir from these water kefir grains. Many of the bacteria and yeast in water kefir grains are the same microbes that inhabit milk kefir grains, but they have adapted to use sucrose and maltose rather than lactose as an energy source. It's possible to wash either type of grain and use it in the other type of kefir, but it may take a while and several batches before the microbes that utilize the new kind of sugar proliferate enough to give the grains real fermenting proficiency in their new surroundings. You'll get a faster water kefir ferment if you

Water kefir grains

start with water kefir grains, either from a friend or your local Fermenters Club, or bought from www.culturesforhealth.com.

For water kefir, I dissolve a couple of tablespoons of Sucanat—natural and organic cane sugar—in 2 or 3 cups of water and add this to a quart jar holding the water kefir grains, which are small, translucent lumps. I add half an organic lemon to give the water kefir a nice citrusy flavor. You can also use just lemon juice or a piece of dried fruit, as long as it's unsulfured. The water kefir sits covered like the milk kefir with paper toweling held tight with a canning lid band in the cupboard next to the milk kefir. After 2 days, or 3 days at most, I drain off the water kefir and store it in the fridge. Now I discard the lemon or dried fruit, then prepare a new batch with the same grains. These grains also reproduce, and, like the milk kefir, extra grains can be frozen. In fact, they reproduce quite easily. When I first got my water kefir grains, they measured about ¼ cup of the translucent blobs. I've already frozen two freezer bags' worth, and the grains in the fermenting quart jar now fill the jar half full—about a pint of grains. It's time to freeze at least half of them.

If you want a particular flavor of water kefir—say black raspberry—cook the berries on the stove gently, until the juice runs, then strain and catch the free juice. Allow this to cool to room temperature and add it to the kefir in the fridge. If you want to have it fizzy, don't put it in the fridge, but cover it with a jar lid screwed down tight with a band and let it sit on the kitchen counter for another day or three, depending on the ambient temperature in your kitchen. The higher the temperature, the faster it will become fizzy. When bubbles have formed, return the jar to the fridge. Besides rendered berry juice, you can add other fruit and herb flavorings by pressing the juice from cherries, grapes, melons, or citrus, or put in grated ginger tied into a cheesecloth bag or a bouquet of mint leaves tied by the stems. Drink this up within a few days because you'll have more water kefir coming and because fresh fruit juices have a relatively short shelf life.

Water kefir can be made with sweetened green tea or fruit juice, but beware commercial, processed fruit flavorings, because these usually contain preservatives that can kill the grains. The sweetening is crucial, because that's what the water kefir grains use as a source of nutrients. The other flavors are just to keep you interested. For a very refreshing draft of water kefir, use coconut water (check to make sure it has no preservatives), sweetening it slightly with Sucanat. This is my favorite kind of water kefir.

Ginger Beer Water Kefir

This is a favorite here and in England. It's said that the water kefir grains and the recipe came home with British soldiers after the Crimean War in the mid–nineteenth century. They called the grains the ginger beer plant. Here's how to make 2 quarts. You can use a glass or ceramic half-gallon jar, as long as it has a tight-sealing lid. A food-grade plastic half-gallon water jug works, too, as long as you have the cap that screws on tightly.

Makes 2 quarts

> 2 ounces fresh ginger root (one thumb)
> ½ cup Sucanat (whole cane sugar)
> 6 cups spring water (not tap water, unless you have your
>   own unchlorinated well)
> 1 cup water kefir grains
> ½ organic lemon
> 1 unsulfured dried fig or 2 tablespoons unsulfured raisins

1. Grate the ginger into a bowl. Sprinkle it with 2 tablespoons of the sugar. With the back of a spoon or pestle, press out as much juice as possible from the ginger, allowing the sugar to dissolve in the juice. Put the contents of the bowl in a double layer of cheesecloth set into another bowl.

2. Draw up the ends of the cheesecloth to make a little bag, tie it off tightly, then twist and squeeze the bag hard, catching the sweetened ginger juice in the bowl.

3. Put the ginger juice and the tied-off cheesecloth bag in the jar with the water and the rest of the sugar and shake or stir until the sugar is dissolved. Add the kefir grains, the lemon, and the dried fruit.

4. Seal the jar tightly and let stand at room temperature for 48 hours, shaking up the contents a few times during that period.

5. Strain off the ginger beer into airtight, sealable, strong plastic or glass bottles. Place in the fridge for 2 days before drinking, or, optionally, increase the fizz by allowing the strained ginger beer to sit sealed at room temperature for an additional day before placing it in the fridge. Drink cold.

Monica Ford's Apple Ginger Soda

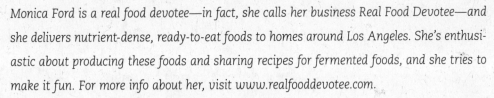

Monica Ford is a real food devotee—in fact, she calls her business Real Food Devotee—and she delivers nutrient-dense, ready-to-eat foods to homes around Los Angeles. She's enthusiastic about producing these foods and sharing recipes for fermented foods, and she tries to make it fun. For more info about her, visit www.realfooddevotee.com.

She starts the process of making fizzing soda by creating what she calls a ginger bug. This is a starter made with fermenting ginger, but she notes you can also use whey, water kefir grains, or kombucha as a starter.

Makes 1 quart

Ginger Bug

> **3 cups filtered water**
> **1 tablespoon Sucanat (whole cane sugar)**
> **1 tablespoon diced unpeeled ginger**

Apple Ginger Soda

> **¼ cup ginger bug**
> **1½ teaspoons freshly juiced ginger (or to taste)**
> **3½ cups freshly juiced apple**

1. Make the ginger bug: Combine all the ginger bug ingredients in a quart jar.

2. Put on the lid, give the jar a shake, and allow it to sit for a week at room temperature. Loosen the lid a few times to make sure too much pressure isn't building up in the jar.

3. Make the soda: Place all the ingredients in a sturdy quart bottle with a tight-fitting screw cap screwed down tight.

4. Allow it to sit in a warm (72 to 80°F) place in your home. If your home is cold at night, use a seedling heat mat or heating pad on the lowest setting and set the bottle on it.

5. After one or two days of fermentation, look for bubbles rising in the bottle. Carefully loosen the screw cap just a bit and listen for the hiss of gas escaping.

6. If you hear the gas, pour the beverage into a glass for a fizzy, probiotic-rich drink. If you'd like it chilled, screw the cap down tight and chill in the fridge until cold, then drink.

Grape Juice Kefir

Take some of your water kefir grains and use them for this delightful beverage. But don't use them all, because the juice will stain them. It tastes best if you are using fresh grapes, especially fresh wine grapes—just make sure they are organically grown, because otherwise they could be sprayed with pesticides or fungicides that can kill off the microbes in your kefir grains.

Serves 4

> **2 bunches fresh organic red or white grapes (3 to 4 pounds)**
> **Spring water to equal the amount of grape juice you get**
> **½ cup Sucanat (whole cane sugar)**
> **¼ cup water kefir grains**
> **½ organic lemon**
> **1 stick cinnamon**

1. Stem the grapes and put the grapes in the blender. Whiz into a coarse slurry on low speed. You want to avoid breaking the bitter, tannic seeds, so blend only until the grapes have turned to mush. Line a bowl with a double layer of cheesecloth and pour the grape slurry into the center of the cloth. Pull up the edges and tie them off to create a bag. Holding the bag above the bowl, twist and squeeze the bag to catch most of the grape juice. Measure the juice amount in a measuring cup.

2. Place the juice in your fermenting jar, which must have a water-tight sealed cap or cover. Add the same amount of spring water as you have juice.

3. Add the Sucanat and stir or shake with the cover on to dissolve completely.

4 Add the kefir grains, the lemon half, and the cinnamon stick.

5. Seal the jar just to finger tight and ferment at room temperature for 1 or 2 days, until fizzy. Then place in the fridge until cold and use within a day. Strain when serving.

OPTIONS: Be creative. Ferment your water kefir beverages with bee pollen, flax seed, or pomegranate juice. Experiment, but remember, water kefir grains need a clean, organic sugar to do their work. Add ½ cup of Sucanat for every half gallon of liquid.

A Well-Hopped Ale

To keep the entire process of beer making under your control means you'd have to grow out a stand of your own barley; harvest and thresh it; sprout the barley, which changes the starch to sugar, at which point it becomes malt; dry the malt; and grind the dried malt. Then you'd be ready to start making beer. Well, you'd be ready if you also simultaneously grew several kinds of hops, gathered Irish moss seaweed at the coast, grew corn to make corn sugar, and tried to find a good strain of beer and ale yeast, which would require the services of a microbiologist. As you can see, while home-brewed beer is certainly possible, it's a lot easier and more fun if you buy your ingredients from a winemaking and beer brewing shop—either a bricks-and-mortar store near you or, even easier, online. A centrally located online source of all the ingredients and equipment used in the accompanying recipe is Midwest Supplies (www.midwestsupplies.com), whose helpful staffers will also answer any questions you may have.

This recipe comes from Peter Burrell, owner and brewmaster at Dempsey's Restaurant and Brewery, perched above the Petaluma River in Petaluma, California. For about twenty years, he's been slaking Sonoma County thirsts with his delicious brews while his wife and co-owner, Bernadette Burrell, has been making sure there's plenty of good food to go with

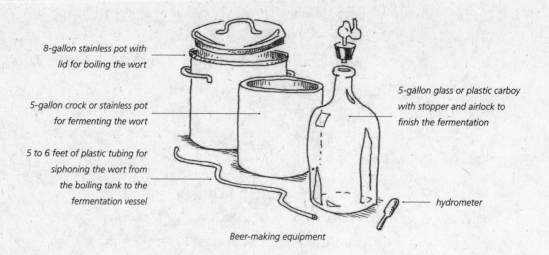

8-gallon stainless pot with
lid for boiling the wort

5-gallon crock or stainless pot
for fermenting the wort

5 to 6 feet of plastic tubing for
siphoning the wort from
the boiling tank to the
fermentation vessel

5-gallon glass or plastic carboy
with stopper and airlock to
finish the fermentation

hydrometer

Beer-making equipment

the perfect pints. Peter says, "This recipe is for an India Pale Ale (IPA), which, as you know, is quite popular these days." Indeed it is. In a recent competition, another Sonoma County IPA, called Pliny the Elder, won the award for best craft beer in America. In describing his recipe, Peter continues, "It's a medium-bodied ale with a hoppy finish that should prove out to be about 6.5 percent alcohol." If you visit Petaluma, be sure to stop by Dempsey's and do a tasting of his beers and ales. They are superb.

Thing one: cleanliness. Keep all your equipment scrupulously clean. It doesn't have to be sterilized, but a good washing and rinsing and maybe a final splash with boiling water are good ideas.

For equipment, you'll need a large—8-gallon at least—metal brew pot with a lid for boiling the wort (the basic brew before fermentation), plus a fermentation vessel like a ceramic crock or large metal pot that will hold 5 gallons of wort. You'll need a 5-gallon glass or food-grade plastic carboy, of the kind used on water coolers. You'll need a rubber stopper with a hole in the center to fit the top of the carboy. A 5- to 6-foot length of plastic tubing will be needed to siphon the liquid from the boiling kettle to the fermentation tank. You'll need an airlock to go into the stopper's center hole in order to keep air away from the young beer before bottling. You'll also need a hydrometer, a device that measures the relative density (what used to be called specific gravity) of a liquid. A scale that measures ounces is useful. And when it comes to bottling, you'll need a funnel, beer bottles (duh), crown cap blanks, and a crown capping device. None of these things is terribly expensive. Here's the recipe:

> 5 pounds dried malt extract (DME), light extract
>
> 2 pounds dried malt extract (DME), amber extract
>
> 1 ounce Columbus hop pellets
>
> 1 ounce Simcoe and 1 ounce Sterling hop pellets
>
> 2 teaspoons Irish moss
>
> ½ ounce Simcoe and ½ ounce Sterling hop pellets
>
> American ale yeast #1272
>
> ½ ounce Simcoe and ½ ounce Columbus hop pellets
>
> ¾ cup priming sugar (corn sugar) for bottling

1. Start with 5 gallons of the best water you can find, such as filtered water or spring water. Put it into the 8-gallon kettle and bring to a boil. Carefully add the malt extracts and 1 ounce of the Columbus hop pellets.

2. After 35 minutes, add 1 ounce each of Simcoe and Sterling hop pellets.

3 At 45 minutes into the boil, add the Irish moss. This seaweed clarifies the wort and helps settle particulate matter.

4. At 60 minutes, add ½ ounce each of the Simcoe and Sterling hop pellets, turn off the heat, and let the kettle's contents settle for 30 minutes. The hops, proteins, and solids will fall to the bottom of the kettle.

5. Carefully siphon the clear liquid from the kettle into the fermentation tank. Use the hydrometer to measure the relative density. It should be about 1062.

6. When the liquid is cool, mix the yeast in a cup of the young beer until it dissolves. Add the yeast slurry to the fermentation tank and stir it in well. Cover the fermentation tank with a clean towel or lid, if it has one. Clean out the brew pot and set it aside.

7. The length of time of the fermentation will vary by temperature, so let the fermentation proceed until there is only minor bubble formation and the hydrometer now reads about 1012. This can take anywhere from 3 to 8 days, depending on the temperature. Let your hydrometer reading be your guide.

8. Add ½ ounce Simcoe and ½ ounce Columbus hop pellets to the carboy, then use the plastic tubing to siphon the fermented wort into the carboy. Fix the rubber stopper with the airlock into the top of the carboy and set the carboy in a cool (68°F is ideal), dark spot for 2 weeks.

9. Carefully siphon the ale into a clean large vessel such as your brew pot, leaving the residual yeast and sediment in the carboy. Add the priming sugar to the ale in the brew pot and stir well until entirely dispersed. Use the plastic tubing to fill beer bottles to about an inch from the top and crown cap the bottles. Put the bottles in cases and set aside for 2 weeks, during which time the corn sugar will re-ferment in the bottles, adding the fizz. For better flavor maturation, wait 2 more weeks before drinking. Chill the bottles well before opening.

Making Kombucha

Ah, the fermented tea we call kombucha—so expensive and fizzilicious from the store, yet so inexpensive to make yourself at home. It's not complicated to make, but it does take some time and can be sort of messy. The trick is to think it through before you set to work so that you avoid mistakes.

Making kombucha is a form of fermentation of sugar to alcohol, performed by the yeast components of the symbiotic combination of bacteria and yeast—the SCOBY. So although you start with a lot of sugar in your sweet tea, the finished product has less than a teaspoon of sugar in a gallon of kombucha and about one-fifth the calories of a can of soda pop.

But doesn't turning all that sugar into alcohol mean that kombucha is strongly alcoholic, like wine or beer? No—because there are also bacteria in that SCOBY, mostly strains of acetic acid bacteria, especially the genus *Acetobacter,* that oxidize the ethanol alcohol to acetic acid, the compound that imparts the familiar tart taste and smell to vinegar. Finished kombucha has between 0.5 and 1.0 percent alcohol. That was enough for the federal government to pull kombucha off the shelves for being mislabeled a while back, but cooler heads prevailed and it was soon returned to the shelves. One would have to drink a prohibitively large amount of kombucha to get an alcohol buzz.

So you start making kombucha by making a sweet tea from sugar and a mixture of black and green teas. Some people wonder about all the caffeine in the black tea, but not to worry. Finished kombucha contains only about a quarter of the caffeine of regular black tea. That's because the many strains of yeast and bacteria that are coexisting happily in the SCOBY exude enzymes that reduce caffeine to harmless substances. The fermentation from sweet tea to kombucha is marked by the major process of sugar to alcohol to acetic acid, but that's just the most salient process. Many other enzymes and biologically active compounds are produced.

Unlike kefir, where we actually drink the healthful bacteria so they can colonize our gut, with kombucha the microbes stay in the mother, or mushroom, or, as we prefer to call it, the SCOBY, and we drink the goodness they create in the sweet tea.

The key to kombucha success is to begin with a healthy, fresh SCOBY. If you have a friend making kombucha, he or she may have one to give you. If not, they are available from many sources, including www.kombuchabrooklyn.com, www.culturesforhealth.com, and www.kombuchakamp.com. Remember that the culture is a living organism, so the younger, more alive, and fresher, the better. Never try to make kombucha from a culture that is dehydrated, has been stored in the fridge or stored in plastic for over thirty days, is moldy, puny, or mushy, or has disintegrating dark spots. It should be firm and healthy-looking, off-white in color, with the consistency of a piece of fresh calamari. Your culture will thrive if it's fed what it likes and is cared for so it doesn't dry out or become neglected. A SCOBY pet? Not a bad way to think about it.

The SCOBY (symbiotic culture of bacteria and yeast), shown here out of its vessel of sweet tea and resting on a platter, is a living organism, so the younger and fresher, the better.

The SCOBY may seem firm and healthy, but it's a somewhat delicate pet. It thrives in glass, ceramic, porcelain, or stainless-steel vessels, but not in plastic—even food-grade plastic—or any other kind of metal vessel, especially aluminum. The pH of kombucha can get down to around 3.0, which is strongly acid, and the acidity can leach toxics from plastic and react with aluminum or any other metal except stainless steel. Don't worry about putting acid in your stomach—it's already full of gastric juice, primarily hydrochloric acid with a pH of 2.0. Your tummy loves acidity. After all, those yogurt-making microbes are called *Lactobacillus acidophilus* (acid-loving) and they like living in your intestines, as well as fermenting your vegetables, krauts, and kimchi.

You can use an organic Sucanat if you wish, but there are elements in those "whole juice" sugars that may slow the fermentation. White, granulated cane sugar is fine, since it is entirely sucrose and is almost all converted during the primary fermentation or secondary fermentation in the bottle (giving the kombucha that lovely fizz). Use no other sweetener for kombucha—that means no honey, no agave sugar, no stevia, and for goodness' sake, no artificial sweeteners like aspartame. The SCOBY wants good old sucrose, period.

For the sweet tea, use only green tea and black tea (but not Earl Grey, which has been doused with the essential oil of bergamot, which your SCOBY will not like). No herb teas, tisanes, or flavorings at this stage. You can add flavorings and herbal decoctions, but only at bottling, after the kombucha has been made. Just be patient and we'll get there.

The only other ingredient in the sweet tea that will become kombucha is some already-made kombucha, the "starter liquid," to make the SCOBY feel at home and encourage it in certain directions. If this is your first batch, borrow some unflavored kombucha from a friend or ask on www.fermentersclub.com if someone near you has some unflavored kombucha to spare. Or you can buy a bottle of unflavored kombucha at the market and pour it into a bowl to allow it to go flat, then use it as a starter liquid. Just make sure it isn't flavored. The bottler may have it labeled "Original," but read the ingredients carefully.

Okay, you have your SCOBY, sugar, filtered water, green and black tea bags (5 or 6 in all), starter liquid, and a proper brewing vessel that will easily hold a gallon or more (a 2-gallon vessel is best). You'll also need a clean dish towel to cover the brewing vessel—no cheesecloth; its weave is too loose and the fruit flies that will be attracted to your vessel as the kombucha is fermenting may be able to get in. A linen dish towel is best, but any tightly woven towel will do. And you'll need a large rubber band or some way to clamp the towel tightly around the vessel. I have used several long twist ties twisted together. They work fine. If you can tie string tightly enough around the vessel, that will work, too.

Your vessels should be freshly washed and perfectly clean, and so should your hands. Some brewers finish hand washing by pouring filtered water over their hands, then drying them on a fresh, clean dish towel or paper towel. Others dip their hands in white vinegar. Whatever. Strive for cleanliness.

After you've set up the brewing vessel with sweet tea, SCOBY, and starter liquid, you will need to find a place that's warm for a proper fermentation to occur. At 75 to 80°F, the fermentation should take from 5 to 7 days. At colder temperatures, it can take 2 weeks, or even stop altogether. You may want to set a heating pad on the lowest setting in a location where you want the brewing to take place, put a large jar or pot of water on it, and take the water temperature after 3 or 4 hours. If this test pot is in the 75 to 80°F range, your brewing vessel will be good to go in that spot. If it's warmer than 80°F, you may have to adjust the position of the test pot above the heating pad—like setting it up on a footstool—until you get the right temperature. With a heating pad, it's unlikely your test pot will be cooler than 70°F, which is low but still okay. Just make sure the heating pad is not covered by anything that could mean a heat buildup and fire.

Homemade Kombucha

1 quart filtered water

¾ cup granulated sugar

5 or 6 green and black tea bags

3½ quarts cool filtered water

2 cups kombucha starter liquid

1 kombucha SCOBY

1. Place the quart of water in a nonaluminum cooking pot and add the sugar. Turn the heat to high and stir occasionally until the sugar dissolves. When the water comes to a boil, reduce the heat to medium, add the tea bags, and boil for 15 minutes.

2. Place the 3½ quarts of cool, filtered water into the brewing vessel. When the sweet tea is ready, let it cool for a few minutes, then remove the tea bags and pour the tea mixture into the brewing vessel. Let the liquid in the vessel return to room temperature. This last is critical, because liquid that's too hot can kill the SCOBY.

3. When the liquid is room temperature, add the kombucha starter liquid and SCOBY to the brewing vessel. The SCOBY may float, or it may sink; this makes no difference.

4. Cover the vessel with the dish towel and clamp it securely to the outside of the vessel.

5. Place the vessel in a warm (75 to 80°F is ideal), airy, dark location. Do not disturb the brewing vessel for at least 5 days. No peeking in to see how it's doing. Leave it alone.

6. Bottle your kombucha following the directions below.

BOTTLING YOUR KOMBUCHA

After 5 to 7 days—the exact timing isn't critical—uncover your brewing vessel and, using a straw, sip some of the fresh kombucha. It should taste and smell a little vinegary or sour and just slightly sweet. Once the batch turns slightly sour, you'll want to bottle it in another day or two. Some sugar left in the kombucha is needed for the fermentation that will continue in the bottle, giving the brew that nice fizz. Try some store-bought kombucha to get an idea of the way you want your batch to taste, but remember, too much residual sugar will mean too much fizz and possibly exploding bottles, so just slightly sweet, please. If it tastes right to you, then it's time to bottle. If it's too sweet, let the fermentation proceed for as many days as it takes for the batch to taste right to you. If it's too sour and vinegary, ferment the next batch for a shorter amount of time.

Let's say you're ready to bottle. You will have been collecting bottles and cleaning them thoroughly for a while. Grolsch beer bottles, with the attached stoppers, are perfect. So are glass kombucha bottles with caps that can be screwed down tightly. Some people use plastic bottles on the theory that if they do explode, they won't send shards of glass flying, but I don't trust the chemicals in the plastic in combination with the acidic kombucha—too much chance of chemicals leaching from the plastic into the drink. Beer bottles are great, but then you'll need to go to your local beer-making shop for crown caps and a bottle capper, or order the same online. Just Google "crown caps and bottle cappers for sale."

A common mistake that people make—especially new brewers—happens at bottling. And that is to mishandle the SCOBY, which, as we've learned, is a rather delicate conglomerate organism. So, before bottling, prepare a batch of sweet tea in a second glass, ceramic, or stainless-steel pot, minus only the SCOBY and the starter liquid. Uncover your brewing vessel. If there's a film on top, that's fine—it's just another "mother" beginning to form. If, however, the film is fuzzy, like the mold you see on old bread, it most likely is mold and you'll want to discard that batch and start again with more sanitary conditions.

Make sure your hands are perfectly clean. Using a plastic strainer—or your clean hands—lift the SCOBY from the brewing pot and immediately place it in the

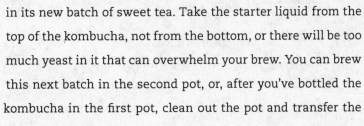

second pot of sweet tea. Now add the starter liquid to the SCOBY in its new batch of sweet tea. Take the starter liquid from the top of the kombucha, not from the bottom, or there will be too much yeast in it that can overwhelm your brew. You can brew this next batch in the second pot, or, after you've bottled the kombucha in the first pot, clean out the pot and transfer the fresh batch back into it.

You may notice some stringy filaments hanging below the SCOBY. These form naturally from the bacteria, yeast, and phenolics in the tea and are not dangerous.

Now is the time to add fruit juice, lemon juice, grated ginger, herb teas, or bits of dried fruit to the bottles to act as flavoring and to give the bottles a little extra sugar for the secondary fermentation that will make the fizz in the bottles. Don't overdo the additions. A teaspoon of juice or fruit will flavor a 12-ounce bottle. Yes, dried fruit and grated ginger will give your kombucha some chewy bits, but they will be wholesome and easily swallowed. Use only 1 teaspoon of liquid flavoring if

WHO IS MAKING KOMBUCHA?

You are providing the conditions, but the SCOBY is actually turning sweet tea into kombucha. And who are the characters at play in the SCOBY? These are the main players, although there may be others.

Bacteria

Acetobacter spp.

Yeast

Saccharomyces cerevisiae—Our old friend who makes our bread and our beer, among many other fermented foods.

Brettanomyces bruxellensis—The bane of winemakers, since Brett, as they call it, gives an unwanted aroma to wine.

Candida stellata—A yeasty fungus that once was classified as a *Saccharomyces*.

Schizosaccharomyces pombe—Much studied, it is a model organism in cellular microbiology. It grows from the ends of its rod-shaped body, then splits in the middle, yielding two cells from one, which is why it's called the fission yeast.

Torulaspora delbrueckii—This is the yeast that imparts banana and clove-like esters to German wheat beers.

Zygosaccharomyces bailii—A spoilage yeast when it inhabits grape juice ferments, yet a beneficial partner when linked to its relatives in a SCOBY.

you want to avoid these extra bits. If you use a charge of white sugar, use only a scant ¼ teaspoon per 12-ounce bottle to avoid too much pressure in your bottles. To bottle, use a plastic funnel and fill the bottles to within ½ inch of the lip, then cap tightly.

Place your bottles in a cardboard case kept at room temperature for 4 or 5 days after bottling, then put them in the fridge. If any bottles do explode, the cardboard case will contain the glass. Open your first bottle in the sink to make sure you don't have too much fizz and soak the ceiling with kombucha.

Some of the kombucha sites recommend a continuous brewing method, using a rather expensive ceramic pot with a spigot. You can check out continuous brewing at www.kombuchakamp.com and at www.getkombucha.com. Both these sites are excellent and contain scads of information, including public posts where kombucha makers ask and answer questions.

Jun

And now, ladies and gentlemen, step into the inner tent of kombucha, the sanctum sanctorum, the holy of holies, where jun resides.

On one level, jun is green tea sweetened with honey and fermented by a jun mother, or SCOBY.

There seems to be a whole other level of jun, very romantic, mysterious, and secret. Emma Blue, writing on www.elephantjournal.com, titled her article "Jun: Nobody Wants Us to Know About It." She says she has an anonymous source who first tasted jun in Tibet, at a camp at the base of Mount Kailash. "The rarest form of jun is Snow Leopard," she says. "The Bonpo monks who produce this fine jun . . . were rumored to have been given heirloom cultures by Lao Tzu.

"The most easily found and tastiest jun in Tibet comes from the Khampa Nomads— former monks turned physical and spiritual warriors who learned how to make jun from the Bonpo. The Khampa Nomads were trained by the CIA in the 1970s to try to kick China out. They took jun so they would have superior fighting abilities against the Chinese. They are also guardians of heirloom cultures, travel on motorcycles with single long braids bouncing off their backs and flasks of jun and swords on their hips."

I don't know if I believe any of this, but someone should make a movie.

And there's more. Jun is made, in secret, it appears, by Herbal Junction

Elixirs in Oregon, because that firm will not divulge the recipe. Ms. Blue reports that people who have tasted Herbal Elixirs jun and Tibetan jun find they taste very similar, but "the jing in Tibetan jun is superior." What is jing? "Jing is the thing that makes you levitate when you've got nothing to lose."

She also says that her family knows a jun dealer who sells jun that he makes himself from a cooler in the back of his car. "Our jun dealer plays gongs for the jun while it brews, as he considers it a living sentient being and it will reprimand you for cursing around it. The first bottle is free."

Again, I can't separate fact from fiction, but I love the idea of an itinerant jun dealer who plays gongs for his ferment. Reminds me of the crazy hippie days.

Here in Sonoma County, the only jun I can find is made in Harbin in Lake County and sold at either Oliver's Market or the Community Market in Santa Rosa. This latter market, originally a health food co-op, has been on the scene for thirty years and stocks whatever you need for your alternative lifestyle. The jun is flavored with elderberry and damiana, this latter being a southwestern and Central American shrub whose flowers have been prized for their aphrodisiac properties, and subsequently banned in certain prudish states. But there's nothing prudish about Harbin Hot Springs, near the jun producer, where those who take the waters usually let it all hang out.

I am not a true believer type. I take a more scientific, skeptical approach to wild claims. But I'll say this about jun: Drinking it makes me want to drink more of it. The elderberry-damiana jun that flows from Harbin in Lake County down here to Sonoma is tasty, fizzy, exciting stuff. Jun—pronounced with the schwa vowel sound, as in the word "won" but with a bit of jing in the sound, somewhere between the sounds in "gin" and "run"—is the beating heart of the kombucha movement.

You can double, triple, quadruple, or in general make more jun, using this as the base recipe.

1 pint commercial jun
5 green tea bags
1 quart filtered water
2 tablespoons honey

1. Pour the commercial jun into a quart canning jar, cover with a piece of paper towel, and screw down the towel with the canning band but not the lid. Set aside in a dark, cool, airy place for 3 to 4 weeks. At the end of this time, a mother should have formed on the jun.

2. Boil the tea bags in the quart of filtered water in a Pyrex or stainless-steel pot for 5 minutes. Remove from the heat and stir in the honey. Allow the liquid to cool to room temperature. That's important. You'll kill the jun mother if you put it into hot liquid. Remove the tea bags and pour the liquid into a brewing vessel of glass, ceramic, or stainless steel.

3. Transfer the jun mother to the brewing vessel and cover with a clean, tightly woven dish towel clamped to the sides of the vessel with a rubber band, string, or what have you.

4. Process exactly for kombucha, tasting after 5 to 7 days—but jun may take a bit longer to become tangy and slightly sweet, since it's a different mix of bacteria and yeast than kombucha's. You may play gongs for the jun mother if you wish, or simply sing it a lullaby. If you have Indian ragas on your iPod, they might be most appropriate. When the jun tastes just right, add a little sweetness so the end product is fizzy. Fruit juice is best, but be restrained—just a tablespoon or two of sweet juice. Too much sweetness means exploding bottles. A few drops of St. Germain elderflower liqueur in each pint would be a nice finish.

5. After it has sat capped for 4 days on the kitchen counter, put the jun into the fridge and drink as necessary. Enjoy the jing.

Kombucha Marinade for Chicken

So you've made too much kombucha and feel rushed to drink it, or you've overfermented a batch and it's really vinegary. What to do? Well, you can preserve its nutritive components and enzymes by using it as a marinade.

Serves 2 or 4

1 thumb fresh ginger, grated

Juice of 1 lemon

1 tablespoon tamari

¼ teaspoon toasted sesame oil

1 teaspoon honey

1 peeled garlic clove

Dash of Sriracha sauce

2 cups fermented kombucha, the culture removed

4 skinless, boneless chicken parts of your choice

1. Place all the ingredients except the kombucha and chicken in a blender and whiz until a thick, homogenous sauce forms.

2. Pour this sauce into the kombucha and stir to mix well.

3. Place the chicken parts in a 1-gallon zip-top freezer bag and add the kombucha marinade.

4. Exclude as much air from the bag as possible and zip it closed. Place the bag in a bowl in the refrigerator for 3 to 8 hours.

5. Take the chicken from the bag, discard the marinade, and cook the pieces any way you wish. I like baking chicken loosely covered with a piece of aluminum foil at 350°F for 50 to 60 minutes, until cooked through, removing the foil at 45 minutes.

Make Your Own Wine at Home

Making wine is one of those fascinating subjects that goes as deep as you care to go. You start by crushing a few grapes and the next thing you know, you're arguing about the effect of pH on the amount of sulfites needed in the must.

Well, forget all that. Like baseball, it's a simple game (you throw the ball, you bat the ball, you catch the ball), even if the rulebook is the size of the Manhattan telephone book. You can make 5 gallons of quite acceptable wine at home with just a little equipment and know-how.

There are wine- and beer-making shops in most large towns, or buy mail-order from one of the many firms with websites accessed by the keywords "home winemaking." And boy, do I have a book for you. *From Vines to Wines* by yours truly will hold your hand through the entire process from choosing a vineyard site to storing the bottled wines.

You'll need:

- A 7- or 8-gallon vitreous crock or food-grade plastic can, either of which should be scrupulously clean
- A clean cotton hand towel, tightly woven
- A 5-gallon carboy (like those used to hold water for water coolers)
- A rubber stopper with a hole in the center that fits the opening in the carboy
- Several clean gallon jugs with 1-hole stoppers to fit their openings
- As many airlocks as you have stoppers
- About 10 feet of ½-inch clear plastic hose used for siphoning
- A large funnel and a mesh plastic bag
- 2 cases (24) absolutely clean wine bottles
- A hand-corking machine and a few dozen corks
- A packet of Campden tablets (sulfites in premeasured form)
- A packet of wine yeast (preferably Prix de Mousse or Champagne yeast)
- 60 to 70 pounds of the sweetest, freshest red grapes you can find, preferably picked from the vine the day you crush them (Avoid Concord, unless you like Mogen David. Seedless table grapes make characterless wine. Look for grapes that have a rich, fruity flavor.)

Place the grapes in the crock or plastic can and pull as many as you can off their stems. Discard the stems. A few stems left in the crock won't hurt. Either tromp the grapes with your (very clean) feet or simply squeeze them with your hands. When they are reduced to a soupy mass and almost all the grape berries are broken open, cover the crock with a clean hand towel and set a board or tray on it to hold the cloth tautly across the top. About 24 hours later, add the correct amount of Campden tablets according to the label directions. Usually they are measured 1 tablet to a gallon. Crush them in ½ cup of warm water until they dissolve, then stir them into the must. About 24 hours later, sprinkle the yeast on top of the must and stir it in. Add winemaker's malolactic bacteria—a Leuconostoc species—and stir it in. It's available from whichever shop sold you your wine yeast. Use the amount proper for the amount of wine you're making. Your wine shop owner will advise you.

At least twice a day from now until you press the wine off the skins, remove the cloth and board and punch down the cap with your hands. Place your hands on the floating cap of skins and push it down under the liquid. Replace the cloth and board after every operation.

In a day or two, the fermentation will be bubbling. It will grow vigorous and then begin to slow down. By about a week, it will either have stopped completely or be bubbling very slowly. Now it's time to separate the wine from the skins. Place the funnel in the clean carboy and line it with the mesh bag. Using a quart jar, scoop up wine and skins from the crock and dump them into the mesh bag so the liquid runs out and down the funnel into the carboy. When the bag is half full of skins, wring it out as hard as you can with your hands to expel as much liquid as possible. The bag's mesh may frequently clog with gunk (spent yeast and grape pulp), so empty it and rinse it clean in the sink when it does. Fill the carboy to within 5 inches of the top, insert the rubber stopper with an airlock inserted into the center hole, and notice whether the wine is still bubbling, sending gas through the airlock. If the wine is still and no gas is escaping, top up the carboy to within 2 inches of the bottom of the stopper. If it's still bubbling, wait until the bubbling ceases (this can take days), then top it up.

Any leftover wine should be put through the mesh-lined funnel into one or more of the gallon jugs. They should similarly be topped up and stoppered with an airlock. If you have a final jug just partially filled, top it up with a good commercial wine similar to the one you're making.

Check the airlocks every few days to make sure they are secure and cannot let air into the carboy or jugs. Store the carboy and jugs in a cool, dark place, such as a cellar. After about 3 or 4 weeks, you'll need to rack the wine off the gunk in the bottom, called the lees. You'll see it as a light-colored layer in the bottom of the carboy or jugs. Without shaking up the wine, carefully set the carboy or jug on a table and place the clean crock on the floor. Using a piece of the plastic hose, siphon the wine from the carboy or jug into the crock. Note carefully where the end of the hose is located in the carboy and avoid siphoning up any of the gunky lees on the bottom. When the clear wine is siphoned off the lees, cover the crock and wash out the carboy. Then pour the clear wine through the funnel back into the carboy. Top up the carboy with wine from one or more of the jugs you racked clean of lees, now to within an inch of the bottom of the stopper and airlock. Consolidate the cleaned wine from the jugs into one or more jugs, topping up the last jug with a similar commercial wine if necessary. Store them again in a cool, dark place. Check the airlocks from time to time to make sure they are keeping the wine safe from contact with the air.

In 3 more months, repeat the racking procedure. This time there will be far fewer lees at the bottom of the carboy or jugs. After this second racking, store them as before, topping up with a sound and similar wine as needed. Check the airlocks periodically.

In 6 months, it's time to bottle. Wet your corks by placing them in a clean container of hot water in which you've dissolved a few Campden tablets. Siphon the wine from the carboy or jugs into each bottle, leaving about ½ inch of space between the level of the wine and where the bottom of the cork will sit when it's placed in the bottle's neck. Place a cork in the hand corker, seat it over the bottle's opening, squeeze the handles together, and with your other hand, plunge the lever downward, seating the cork. It'll take a few bottles before you get the hang of it. Have a corkscrew on hand to remove any corks that break or don't sit right. If there is no airspace between the wine in the neck and the cork, or too much space, withdraw the cork with the corkscrew, adjust the level, and recork with a fresh cork.

Place the freshly corked bottles in their cases, take them to your cool, dark storage area, and store the cases on their sides, so the bottles are horizontal and the wine is in contact with the cork. Wine should always be stored like this. The optimum storage temperature is an even 58°F, but don't worry if it's 70°F in the back of your clothes closet. A steady, even, and not too warm temperature will do.

At some point during the next year, make labels and put them on the bottles with a glue stick. Wine- and beer-making shops will have labels and foil capsules. You can make the bottles look as fancy as you please. Just don't put fancy labels on bad wine, or you will look pompous and silly.

Let the wine age for a year after bottling. What will it be like? One never knows, but surprises are usually more pleasant than not.

Resources

Local Artisanal Cheeses of Sonoma County

Local for me means Sonoma County, California, with its westernmost third composed of about five hundred square miles of lush, parklike grass and oak hills bordering the cold Pacific Ocean. The hills here and in the contiguous northern reaches of Marin County turn green each winter and spring with a gorgeous display of annual grasses and colorful wildflowers. It used to be those hills were thought too cold for growing wine grapes, and ranchers ran cattle and milk cows on them. Starting in the mid-1980s, sheep joined the cows and local sheep's milk cheeses began to be made. By the 1990s, vintners and grape growers discovered that the cool hilltops grew world-class Pinot Noir. Today, those wonderful fermented partners, cheese and wine, flow in abundance from the west Sonoma and north Marin hills.

National Geographic recently named the ten best places to visit around the world.

Only one place in the United States was included: Sonoma County. It's a tourist destination for many good reasons, including stunning natural beauty, great bakeries, wineries, and cheesemakers. If you plan on visiting this beautiful and fecund place, copy the following list of fromageries and be sure to do some serious sampling while you're here.

BLESSED ARE THE CHEESEMAKERS

Availability of these producers' products varies. Check the websites for details. All are within an hour's drive from Santa Rosa, Sonoma County's centrally located main city, and not much farther from San Francisco.

Achadinha Cheese Company. Good goat cheese since 1955. www.achadinha.com

Andante Dairy. Some of the region's—even the country's—best cheeses come from Soyoung Scanlon's operation. www.andantedairy.com

Barinaga Ranch. Marcia Barinaga's ancestors were Basque shepherds and she makes her Txiki from raw sheep's milk. www.barinagaranch.com

Bellwether Farms. The Callahans' sheep's and cow's milk cheeses are superb. www.bellwetherfarms.com

Bleating Heart. Sheep's milk cheese in spring and summer, cow's milk Sonoma Toma in fall and winter. www.bleatingheart.com

Bodega Artisan Cheese. A man from Peru makes the purest and silkiest goat cheese imaginable. www.bodegaartisancheese.com

Bohemian Creamery. Époisses-like sheep's milk cheese is seasonally available and much sought after. www.bohemiancreamery.com

Cowgirl Creamery. Two gals in Point Reyes on the coast make luscious double and triple cream cheeses. www.cowgirlcreamery.com

Laura Chenel's Chèvre. Laura was one of those who started the craze for great Sonoma goat cheese. www.laurachenel.com

Marin French Cheese. Cheeses similar to Camembert and Brie. www.marin frenchcheese.com

Matos Cheese Factory. No sign on the road. (707) 584-5283. No credit cards. Best to pick up Joe Matos's excellent Portuguese-style cheese at a market.

North Bay Curds and Whey. Alissa Shethar uses raw sheep's and cow's milk to make a variety of fresh and aged cheeses. www.northbaycheese.net

Point Reyes Farmstead. Among the best blue cheeses anywhere. www.pointreyes cheese.com

Ramini Mozzarella. True water buffalo mozzarella from local animals is a new addition to the cheese scene. www.raminimozzarella.com

Redwood Hill Farm and Creamery. Artisan goat cheese, goat yogurt, goat kefir—you get the idea. www.redwoodhill.com

Spring Hill Jersey Cheese Company. Sells a variety of artisan cheeses and cultured butter. www.springhillcheese.com

Toluma Farms. A goat ranch in Tomales, Marin County, bordering Sonoma. Fresh and aged goat cheese. www.tolumafarms.com

Two Rock Valley Goat Cheese. The Italian-Swiss heritage of the owners shows in their aged and fresh goat's milk cheeses. (707) 762-6182

Valley Ford Cheese Company. Estero Gold is a brisk, nutty, Asiago-like cow's milk cheese. www.valleyfordcheeseco.com

Vella Cheese Company. One of Sonoma County's oldest fromageries, known for its Italian-style Stravecchio cheese. www.vellacheese.com

Weirauch Farm and Creamery. Nicely aged sheep's and cow's milk cheeses from their site east of Petaluma. www.weirauchfarm.com

Books

Dunn, Rob. *The Wild Life of Our Bodies.* New York: HarperCollins, 2011.

Gasteiger, Daniel. *Yes You Can! And Freeze and Dry It, Too.* Brentwood, TN: Cool Springs Press, 2011.

Katz, Sandor Ellix. *The Art of Fermentation.* White River Junction, VT: Chelsea Green Publishing, 2012.

———. *Wild Fermentation.* White River Junction, VT: Chelsea Green Publishing, 2003.

Lipski, Elizabeth. *Digestive Wellness,* 4th ed. New York: McGraw-Hill, 2012.

Websites

www.BakersCatalog.com · Everything for the home bread baker from King Arthur Flour, P.O. Box 876, Norwich, VT 05055.

www.bodyecology.com · Learn more about how you and your microbes work together to improve your health.

www.culturedpickleshop.com · 800 Bancroft Way, Suite 105, Berkeley, CA 94710

www.culturesforhealth.com · Living cultures and equipment for making a wide range of fermented foods at home.

www.dairyconnection.com · Everything you'll need to make cheese at home.

www.fermentersclub.com · The Fermenters Club shares recipes and expertise, equipment and techniques. A must-visit website.

www.freestonefermentationfestival.com · The fermentation festival is an annual event, held now at the Sonoma County Fairgrounds in September (it outgrew little Freestone, a quaint village in the western part of Sonoma County).

www.getkombucha.com · Great information about continuous brewing and equipment for sale.

www.healingspringsdrinks.com · Jun for sale. Jun is made by Healing Springs Herbal Drinks, 18424 Harbin Springs Road, Middletown, CA 95461.

www.kombuchabrooklyn.com · SCOBY for sale, fermenting vessels, and more.

www.kombuchakamp.com/kombucha-cultures · Another source for fresh cultures to make kombucha and jun, and kombucha-making equipment.

www.mamakai.org · MamaKai provides nourishment and support for growing families in the Bay Area. Home delivery of breakfasts, lunches, dinners, snacks, and beverages made from local organic, biodynamic, and pasture-raised ingredients. Telephone (510) 325-4785.

www.midwestsupplies.com · Whatever the home brewer of beer and ale needs, this place has.

www.myspicesage.com · A source for teff flour for making injera, a fermented Ethiopian bread.

www.nourishedkitchen.com · A good-looking website with lots of recipes including fermented foods.

www.rareseeds.com · Grow your own cucumbers from this catalog, then make your own fermented pickles.

www.redboatfishsauce.com · Make organic kimchi at home with a bottle of this fish sauce for flavoring.

www.southrivermiso.com/store/p/13-Organic-Brown-Rice-Koji.html · A source for koji rice to make your own miso.

www.yemoos.com · Sells milk and water kefir grains and supplies. I ordered my kefir grains here and am well pleased.

www.yourwildlife.org · A great website, especially for young folks, dedicated to "exploring the wildlife that lives on us, in us, and around us."

References

Aris, A., and S. Leblanc. "Maternal and Fetal Exposure to Pesticides Associated to Genetically Modified Foods in Eastern Townships of Quebec, Canada." *Reproductive Toxicology* 31, no. 4 (May 2011): 528–33.

Bisson, L. F. "Biotechnological Modification of *Saccharomyces cerevisiae*: Strategies for the Enhancement of Wine Quality." In Jean-Richard Neeser and J. Bruce German, eds., *Bioprocesses and Biotechnology for Functional Foods and Nutraceuticals* (New York: Marcel Dekker, 2004): 68–87.

Fierer, N., M. Hamady, C. L. Lauber, and R. Knight. "The Influence of Sex, Handedness, and Washing on the Diversity of Hand Surface Bacteria." *Proceedings of the National Academy of Sciences USA* 105, no. 46 (November 18, 2008): 17994–99.

Gill, S. R., M. Pop, R. T. Deboy, P. B. Eckburg, P. J. Turnbaugh, et al. "Metagenomic Analysis of the Distal Gut Microbiome." *Science* 312, no. 5778 (June 2, 2006): 1355–59.

Ley, R. E., C. A. Lozupone, M. Hamady, R. Knight, and J. I. Gordon. "Worlds Within

Worlds: Evolution of the Vertebrate Gut Microbiota." *Nature Reviews Microbiology* 6 (2008b): 776–88.

Ley, R. E., P. J. Turnbaugh, S. Klein, J. I. Gordon, et al. "Human Gut Microbes Associated with Obesity." *Nature* 444, no. 7122 (December 21, 2006): 1022–23.

Li, M., B. Wang, M. Zhang, M. Rantalainen, S. Wang, H. Zhou, et al. "Symbiotic Gut Microbes Modulate Human Metabolic Phenotypes." *Proceedings of the National Academy of Sciences USA* 105, no. 6 (February 12, 2008): 2117–23.

Marcy, Y., C. Ouverney, E. M. Bik, T. Lösekann, N. Ivanova, et. al. "Dissecting Biological 'Dark Matter' with Single Cell Genetic Analysis of Rare and Uncultivated TM7 Microbes from the Human Mouth." *Proceedings of the National Academy of Sciences USA* 104, no. 29 (July 17, 2007): 11889–94.

Mira, A., R. Rushker, and F. Rodriguez-Valera. "The Neolithic Revolution of Bacterial Genomes." *Trends in Microbiology* 14, no. 5 (May 2006): 200–206.

Rawls, J. F., et. al. "Reciprocal Gut Microbiota Transplants from Zebrafish and Mice to Germ-Free Recipients Reveal Host Habitat Selection." *Cell* 127, no. 2 (October 20, 2006): 423–33.

Reganold, J. P., P. K. Andrews, J. R. Reeve, L. Carpenter-Boggs, et al. "Fruit and Soil Quality of Organic and Conventional Strawberry Agroecosystems." *PLOS ONE* 5, no. 9 (2010): e12346.

Savage, D.C. "Microbial Ecology of the Gastrointestinal Tract." *Annual Review of Microbiology* 31 (1997): 107–33.

Turnbaugh, P. J., et al. "A Core Gut Microbiome in Obese and Lean Twins." *Nature* 457 (January 22, 2009): 480–84.

Turnbaugh, P. J., R. E. Ley, M. Hamady, C. M. Fraser-Liggett, R. Knight, and J. I. Gordon. "The Human Microbiome Project." *Nature* 449 (October 18, 2007): 804–10.

Turnbaugh, P. J., R. E. Ley, M. A. Mahowald, V. Magrini, E. R. Mardis, and J. I. Gordon. "An Obesity-Associated Gut Microbiome with Increased Capacity for Energy Harvest." *Nature* 444 (December 21, 2006): 1027–31.

Index